# MISTAKEN IDENTITY

SUNY series in Science, Technology, and Society
Jennifer Croissant and Sal Restivo, Editors

# MISTAKEN IDENTITY

## THE MIND-BRAIN PROBLEM RECONSIDERED

LESLIE BROTHERS

State University of New York Press

Published by
State University of New York Press

© 2001 State University of New York

For information, address the State University of New York Press,
90 State Street, Suite 700, Albany New York 12207

Cover design by Jon Hofferman

Marketing by Fran Keneston • Production by Bernadine Dawes

**Library of Congress Cataloging-in-Publication Data**

Brothers, Leslie.
    Mistaken identity: the mind-brain problem reconsidered / Leslie Brothers.
    p. cm.—(SUNY series in science, technology, and society)
    Includes bibliographical references (p. ) and index.
    ISBN 0-7914-5187-9 — ISBN 0-7914-5188-7 (pbk.)
    1. Mind and body.  2. Philosophy of mind.  3. Cognitive neuroscience
4. Cognition—Social aspects.  I. Title.  II. Series.
BF161 .B763 2002
128'.2—dc21                                           2001020143

1   2   3   4   5   6   7   8   9   10

For Sue and Dwight

# Contents

# Preface

I offer this book as a contribution to brain science, for, as Joseph Rouse (1996) wrote, we should "expand the conception of 'science' to include critical reflection on its own practices and goals"(1996, 31). This is a work of critical reflection. It is written for anyone interested in stories about the brain being the basis of mind. The reader might be a layperson interested in psychology and human nature; he or she might be someone who works in an academic field such as neuroscience, psychology, or sociology. But no prior experience is required.

Chapter 1 explains the mind-brain problem and its various proposed solutions in simple, outline form. Chapter 2 takes the most popular of these solutions, the idea that mind somehow arises from the individual brain, and analyzes it along lines pioneered by the philosopher Ludwig Wittgenstein. There we begin to understand why cognitive neuroscience's current direction is fundamentally mistaken.

We are then obliged to confront an apparently successful and growing literature that purports to show how the mind arises from the brain. So in the succeeding four chapters, we take specimens of this literature and dissect them. What we find are a number of devices for covering over a basic logical problem, none of which work.

Chapter 7 surveys various social agendas covertly served by the popular, brain-based approach. Chapter 8 looks at the actual state of neuroscience in its relation to psychology and concludes that psychology, not neuroscience, is running the show. Chapter 9 suggests a different role for neuroscience in the mind-brain problem. It reviews findings from the nascent area of social neuroscience that reveal the brain's exquisite sensitivity to social gestures. Building on this foundation, chapter 10 offers an alternative solution to the mind-brain problem that respects both our everyday ways of talking about the mind and the neural matter inside our skulls.

I would like to acknowledge the gifted colleagues and friends who contributed directly and indirectly to this work. I thank Sal Restivo for many instructive conversations and general intellectual cama- raderie. Nancey Murphy was the first to encourage me when I was trying out the ideas here. She subsequently provided valuable sug- gestions for the expository structure of the book and guided me to essential works. I am most grateful to her. Jon Hofferman gener- ously spent long hours both on graphics and editing, greatly im- proving the book's readability. I would also like to thank my father, Dwight Brothers, for helpful suggestions regarding content.

My dear friends Jochen Braun and Naomi Wolman galvanized me into action when the manuscript was in danger of collecting dust indefinitely. It is literally true that this work would not have ap- peared had it not been for them.

No one mentioned here is in any way responsible for the book's flaws, however. I alone am responsible for those.

# 1

# The Mind-Brain Problem

This book is about a popular solution to the mind-brain problem. Therefore it may be helpful to begin with what we mean by the "mind-brain problem."[1] In essence, the question is, What is the relation between our subjective experiences and our physical bodies?

This relation is puzzling for several reasons. First, it is hard to explain how something that exists in space (the body) can contain things that aren't physically bounded, like thoughts and beliefs. Second, how can the way it *feels* to have an experience be based on *physical* stuff (like the brain)? And third, it is extremely difficult to explain how mental things like reasons get translated over into the physics and chemistry of bodily movements—and we'd have to explain that to account for a person *doing* something physical because of a mental thing like a belief or desire.

Many serious thinkers have struggled with these issues at length, trying to come up with solutions. In general, their solutions fall into the following categories.

1. Dualism, which holds that mind is something altogether separate from the natural world. This position implies two separate kinds of worlds, the natural and the extranatural, and is rejected by most scientifically oriented people today.

2. Naturalism, which holds that the mind belongs to the natural world.

a. Mental states are forms of brain activity and don't have any real properties of their own. People holding this

1

view are called "eliminativists." They think everyday psychological language is just outmoded talk that will one day be replaced by the more accurate language of neuroscience. Or,

   b. Mental states are aspects of brain states—like liquidity is a property of water molecules at certain temperatures—with bona fide properties of their own, including the ability to cause physical events. Or,

   c. Mental states are aspects of brain states, but without causal properties of their own.

Naturalist philosophers and neuroscientists generally have an *internalist* view of the mind, meaning they think it has to do with the individual rather than society. Specifically, they tend to agree that the mind depends on the individual brain. But they argue quite a lot about positions a–c.

   3. Ordinary language philosophy

   a. Mental states are a way we have of talking about ourselves, but do not have any reality apart from behavior (including talk).

   b. Mental states are tied to social practices: thoughts and words get their meanings from the ways they are used in forms of life in which the thinker or speaker participates. A variant holds that mental states are tied to social practices because thoughts are merely imagined verbal responses to a situation.

These approaches are more popular among sociologists than among neuroscientists and most psychologists. They can be called *externalist* since they locate the mind in a network of practices, rather than internally in an individual.

A final possibility is that there is no solution to the mind-brain problem. Holding on to naturalism and rejecting dualism, we may be suspended in an "explanatory gap" between the mind and the brain. Many scientists and philosophers think this is exactly where we are—and of course they are not satisfied. Increasingly, especially among eliminativists, there is doubt that any amount of philosophy will close the gap. Instead, in these and other quarters, there is growing confidence that advances in neuroscience will finally solve the mind-brain problem, and solve it in naturalist, internalist terms.[2]

## NEUROISM

The position that the mind can be explained in terms of the individual brain I will label "neuroism." Neuroism is very pervasive and expressed in a range of sources, from scientific journal articles to popular works in magazines, books, and newspapers.

There are several problems with neuroism. One is that it just doesn't make sense. Most of this book is devoted to showing that neuroism is a logical failure. The basic logical issue is outlined in chapter 2. Chapters 3 through 6 provide examples of the flaws found over and over again in neuroist stories: these indirectly reveal the core logical mistake. In addition, a purely practical reason why neuroism doesn't work is given in chapter 8.

The second problem is the role neuroism plays in our lives. Neuroism *naturalizes* ideology: it makes our culturally derived ideas about ourselves seem to come from nature—to be facts about the brain. It thus allows the social arrangements behind ideologies of the human mind to remain hidden. (This function of neuroism and some of the interests behind it, the use of neuroist narratives as status emblems, and the cultlike features of neuroism, are all developed in chapter 7.) Thus, as a guide to who we are, neuroism is not only wrong, but it is also deeply misleading.

## NEUROSCIENCE, COGNITIVE NEUROSCIENCE, AND NEUROISM

Neuroscience is not inherently neuroistic—that is, not all neuroscientists are interested in showing that the individual brain is the basis of the mind. The field is tremendously broad, extending from studies of the neuronal membrane to topics such as how information is encoded by populations of neurons. Many neuroscientists simply want to understand how the brain works. What causes a neuron to generate an action potential? How does the nervous system develop and become specialized for various functions? How do the signals passed to the brain from the eyes allow us to see? Like scientists in general, they want to understand more about the natural world.

Neuroscientists also search for solutions to illnesses such as Huntington's disease, multiple sclerosis, stroke, schizophrenia, and Alzheimer's disease.

Cognitive neuroscience is the subdiscipline of neuroscience that attempts to bridge the brain and psychology. (We shall use "psychology" to mean the way the mind is supposed to work, so "psychology" and "mind" are practically synonymous.) One of its distinguished proponents defines cognitive neuroscience as the exploration of how primary neurobiological data "speak to the issues of how brain enables mind" (Gazzaniga 1995, xiii). It is in cognitive neuroscience, then, that we find the neuroist program in play.

## THE EXCITEMENT IN COGNITIVE NEUROSCIENCE

The biologist E. O. Wilson recently captured cognitive neuroscience's spirit of bold self-confidence when he wrote that its characteristics "are the hallmark of the heroic period, or romantic period as it is often called, experienced by every successful scientific discipline during its youth. For a relatively brief period, rarely more than half a century, researchers are intoxicated with a mix of the newly discovered and the imaginable unknown." During this phase, he continued, researchers

> bear comparison with explorers of the sixteenth century, who, having discovered a new coastline, worked rivers up to the fall line, drew crude maps, and commuted home to beg for more expeditionary funds. And governmental and private patrons of the brain scientists, like royal geographic commissions of past centuries, are generous. They know that history can be made by a single sighting of coastland, where inland lies virgin land and the future lineaments of empire. (1998, 100)

Wilson believes cognitive neuroscience is truly in a heroic period similar to those he himself has witnessed in molecular biology, geology with its discovery of plate tectonics, and what is termed the "modern synthesis" in evolutionary biology.

Cognitive neuroscience is unquestionably *feeling* heroic: the inflow of resources, as Wilson suggests, is remarkable, with universities

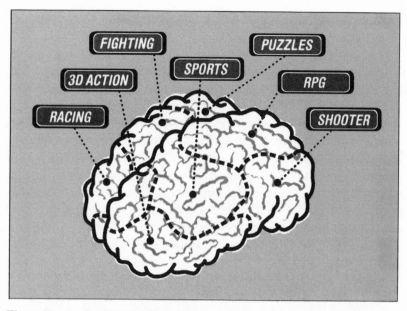

Fig. 1. Illustration from video game promotional pamphlet, courtesy of Sony Electronics.

and research institutes scrambling to build expensive brain imaging centers and hire new faculty in this area. The growth in sheer numbers of neuroscientists has been exponential in recent decades, as membership in the Society for Neuroscience and the sparkling success of the more recent Society for Cognitive Neuroscience attest.

The idea that psychological questions will be answered by neuroscience has captured the popular imagination too. For example, an article heading the sports section of the *Los Angeles Times* (9 August 1998), entitled "Brain Matters," profiled the work of sports consultant Jonathan P. Niednagel, nicknamed the "Brain Doctor:"

> Niednagel is not a scientist, but has combined years of research and a passion for sports to produce "Your Key to Sports Success," a book in which he "types" athletes and other luminaries. . . . Niednagel says [Michael] Jordan's rare combination of physical and mental superiority has made him an athlete for the ages. . . .
>
> "I could show you a picture of Michael's brain," Niednagel says. "He's strongest in the right posterior part. His No. 1 gift is spatial thinking. It's his No. 1 cognitive trait.". . .

Niednagel charges $500 a head but says that brain typing will one day be done with a blood test.

The article is accompanied by a picture of the "Brain Doctor" gazing out from a backdrop of brain scans and anatomical charts.

Apparently there are pretty fair financial rewards for using cognitive neuroscience terminology to persuade people one can predict athletic performance. More seriously, the need for explanations of how we think and remember, why some children disrupt their classrooms, why some people are addicted to drugs, and how we feel fear or joy is generally compelling. Society has a keen interest in and avid appetite for brain-based explanations of human behavior. Thus, amid much fanfare, Congress declared the decade stretching from 1990 to 2000 to be the Decade of the Brain. Outside the walls of academia, images of the brain appear more and more frequently, for example in advertisements such as that shown in figure 1.

The trouble is that neuroscience's sense of exuberance does not prove its mission is succeeding, or even going in the right direction. Several decades of enthusiasm for neuroscience's promise to explain the mind doesn't necessarily mean that there is a scientific revolution, such as those cited by Wilson in molecular biology, plate tectonics, and evolutionary theory, under way. In fact, as we shall see, there are significant cracks beneath the glossy surface of neuroscience's declarations of imminent triumph over the mind-brain problem, cracks that have been masked by the enthusiasm of both neuroscientists and the larger public.

# 2

## Pictures

In the last chapter, we saw there is great confidence that the mind will be explained in terms of the brain. Although many are sure neuroscience will provide the answers, there is a very wide explanatory gap between the brain and the mind that does not seem to be narrowing. The real nature of the gap will become clear in this chapter. We will see that neuroism has a special relation to our ordinary ways of talking about ourselves and the world, a relation that makes it logically impossible for neuroism to solve the mind-brain problem.

The philosopher Ludwig Wittgenstein (1958) showed that in our everyday lives we have complicated and often ambiguous ways of talking about mental things. He called our actual uses of mental terms (like "feel" or "understand") the *grammars* of those concepts. While the word may sound like it has to do with nouns and verbs, he was using it in a more general way. He meant to denote the complex ways we talk about things in ordinary conversation. In the case of mental terms and other everyday concepts, we learn how to use the terms appropriately—that is, how to speak within the pattern of accepted practice—as we grow up and participate more and more in society and its conversations. Wittgenstein's "grammar" simply means "the way one ordinarily talks about something."

Even though the actual uses of mental terms are complicated and subtle, speakers and listeners in everyday situations don't get bogged down about what they mean. Instead, by jointly taking much for granted, they communicate well enough for the practical purposes of the moment, and are able to keep on going. So as long as we use mental language in the everyday situations where it naturally belongs, we do fine.

However, when we begin to scrutinize mental concepts such as feeling or understanding, instead of just using them, trouble begins. We tend to make pictures of what the words might mean, pictures that are much simpler than their complex, everyday uses. We begin to see the picture as a more satisfying version of the concept than the speaking practices that gave rise to it in the first place. As Wittgenstein (1958) wrote,

> A picture is conjured up which seems to fix the sense *unambiguously*. The actual use, compared with that suggested by the picture, seems like something muddied.
>
> We are tempted to say that our way of speaking does not describe the facts as they really are. (1958, 127, 122)

As a result of our scrutinizing and picturing, we begin to think that the ordinary ways of talking *obscure* the concept. We get the notion that the actual nature of mental terms (like "remembering" or "paying attention" or "being depressed") is somehow hidden beneath the indeterminacy of our language practices. Everyday ways of talking, for example about emotion or pain, begin to be seen— mistakenly—as not-quite-exact accounts of the real phenomena.[1]

Now let's turn to the pictures themselves. Our pictures of mental concepts are meant to capture, in a kind of shorthand, the complex ways we have of speaking of them. In the case of subjective experience, for example, we use the picture of inner experience as opposed to the outer world. In picturing what we mean by "feeling," we imagine pointing to something inside ourselves with our attention or our imagination, just as we might point to something with our finger when we consider an object in the world. Wittgenstein said such pictures should be taken as "grammatical remarks"—simplified renderings of the specialized areas of our language having to do with mental concepts.

When the pictures are taken as more than they really are—as telling us something real about the concepts—they cease being grammatical remarks or pictures, and become illusions. Philosophers and psychologists muddle themselves by trying to enter ever more deeply into such pictures. To clear up their muddles about mental states, Wittgenstein recommended going back to the everyday grammars. As the reader has guessed or is already aware, Wittgenstein is one of the originators of the "ordinary language" approach to the mind mentioned in chapter 1.

Now we are ready to show what is wrong with neuroism. The neuroist notion that the mind can be found in the individual brain is an illusion based on a picture of the human person. It is a *single* picture of both physical body and mental life. As a grammatical remark, this picture accurately captures the way we use the term "person" in normal, practical circumstances, for "a body with a mental life" is exactly what we mean by "person" (Strawson 1959). But when we move away from our everyday use of "person" to scrutinize the picture more deeply—when we start to take the picture as an object of analysis—we get confused. We feel we should bring to bear *both* the language of natural science (for the physical part of the picture) *and* the language of the mental (for the psychological part of the picture) *to describe the same picture.*

The trouble is, the practices of modern natural science belong to one area of human thought, with its own history and concepts, while the practices of attributing mental states to ourselves and others belong to an entirely different area, also with its own history and concepts. As Wittgenstein might have put it, science on the one hand, and everyday psychology on the other, are two separate language-games. So even though we normally fuse body and mind in the concept of "person," it doesn't work to take the picture so seriously that we drag in the full conceptual paraphernalia of natural science in order to weld it together with psychology.[2]

At this point the reader may object, well, isn't psychology a science too? Are they really so different? It is a common confusion, not least on the part of psychologists themselves, to think psychology is a science. So let's say more about what we mean by "science," and show why psychology is fundamentally different.

## TWO LANGUAGES

Scientific practice involves ideas of true and false. It involves the idea of natural laws that scientists can't know directly but can learn about through experiments. In scientific practice, facts and data are communicated in a way that makes it clear they are meant to refer to something observable and objective. Neuroscience research adheres to this set of practices and this language.[3]

In contrast, concepts of the mind do not ultimately refer to observables. When psychologists study entities such as aggression, depression, or self-esteem, they have to attribute motives to their subjects in order to interpret the actions they observe. Therefore, the definition of what's being measured in a psychological "experiment" must itself be framed using ideas about other, underlying psychological dispositions. As a result, statements about psychology are never ultimately anchored in observables; instead, they are always anchored in other psychological statements.[4]

What experiments in psychology actually do is creatively reencode concepts we already have, not test theories by rejecting false hypotheses, as noted by psychologist John Shotter: "[Psychology] cannot, as it hopes, penetrate into men's actual inner 'workings,' for it can never escape from the fact that we only ever investigate what we mean by the term 'our inner workings.' And while the meaning of the term may be changed as a result of our investigations, even then we still confront *what we mean by the term* rather than our actual inner workings 'in themselves'" (1975, 26; italics added). As Shotter points out, psychology attempts to blur the distinction between the discourse of mental life and the standard discourse of science. It holds itself to be doing science when it is really doing image-making—that is, elaborating the language of the mental to give us new pictures of ourselves.

The upshot of these considerations is that the truth of a statement about something mental is actually not an empirical matter. It's simply a matter of appropriate or inappropriate *use* of mentalist language, in a particular circumstance, as sanctioned by the speaking community. Ultimately, the language of the mental is anchored in society.

There are good reasons for assenting to natural scientists' claims that they make contact with an external reality. Certainly, the forms their knowledge takes have much to do with their own collective social practices (which they themselves disregard). However, to deny that science engages with the external natural world is an extreme position. By contrast, to deny that even so-called scientific psychology engages with an objective, external world is not an extreme position. As a set of practices, "scientific" psychology is hardly distinguishable from everyday psychological talk, whereas the natural sciences are highly distinguishable from folk theories of how the world works. However disguised, this is a key difference between the terrain of mental theorizing and the terrain of science.

There are other, less subtle, differences between science and psychology. Psychological concepts are ancient. They are organized around our notions of the person, and primarily concern such things as motivations, beliefs, and accountability. Modern science—the science we know—arose only about three centuries ago, and is organized around the ideas of true and false, replicability, and an independent objective world. Taken together, all these features show that science and psychology are separate areas of organized human practice.

Forgetting that psychology and science are different kinds of social practices leads to the misguided attempt to apply both to the same picture. True, our use of the person idea in everyday language does lead to a single picture, but the picture is just a grammatical remark, not something meant to be deeply scrutinized. It should not seduce us into forgetting that the languages of natural science—including neuroscience—and of the mind are fundamentally separate. Forgetting this, and taking the picture as more real than it is, are the twin sources of the problem in the mind-brain problem. (Logically, of course, we could avoid forcing some kind of equivalence between psychology and science by denying the mind exists. This would save us from a confused approach to the everyday picture of person, but only at the expense of confusing us in our everyday world. This option is discussed at greater length in chapter 9.)

## EVIDENCE OF INTRACTABILITY

The argument so far is that taking the everyday notion of person as a picture to be penetrated deeply, and analyzing it as a single picture using *both* scientific and psychological concepts, is logically confused. If the argument is correct, then the dominant approach to the mind-brain problem is in a hopeless mess.

There are two kinds of evidence that intellectuals are indeed in a deep, intractable muddle over the mind-brain problem. One is that they behave as though they are just about worn out, as though the picture has become like the tar baby in the Br'er Rabbit story—exhausting to tangle with. For only exhaustion or desperation can explain why a sizable group of philosophers has agreed to hand the problem over to the neuroscientists.

But maybe the handover is working. Aren't there now many books, very positively received and held to be illuminating, that tell us exactly how the brain produces the mind? Well, yes—there are many. But when we look at these efforts, we uncover the second kind of evidence of a deep muddle. It is that actual neuroist theorizing is jury-rigged through and through. In the next four chapters, we look over neuroist accounts themselves. We will soon be convinced that the handover is not working at all.

One could object that it's just too early for the results that neuroism promises. If the fit between brain language and mind language isn't perfect now, one could argue, in due time our concepts of mind will be refined so that they dovetail with what we are learning about the brain, and neuroism will work. We shift the grounds of this objection by asking the reader to focus throughout the next few chapters not on what neuroism *might* do, but on what it *does* do. If neuroism is to refute the argument spelled out in this chapter, it will be in the form of particular, successful accounts. In our upcoming analyses of currently respected accounts, we won't be concerned at all with how closely concepts like attention, emotion, or consciousness fit the neurological facts, for these are by no means final. What we will determine, instead, is whether particular neuroist arguments are internally logical and coherent. If none of them is *logically* sound, there must be a more significant obstacle than insufficient experimental data.

## OVERVIEW OF NEUROIST WRITING

The common theme of the next four chapters is that all neuroist writing faces the same dilemma—namely, how to patch the separate discourses of neuroscience and psychology together. On the surface, this patching looks easy—even amateurs can do it. This is because neuroscience still lacks a central theory and is therefore quite permissive as to how its data can be handled and interpreted. (In-depth discussion of this feature of neuroscience is deferred to chapter 8.) But, regardless of how easy it may be to assemble neuroscientific facts into interesting narratives, the underlying illogic isn't erased. The ineradicable gap between the discourses of brain and mind remains visible, once we learn to see it, through the cracks and at the edges of neuroist narratives.

There are three main neuroistic approaches to the link between brain-talk and mind-talk. The first approach, surveyed in chapter 3, blithely takes them as the same kind of talk. In this approach, some superficial similarity between a psychological term and a neuroscience term may be exploited by ignoring the very different contexts in which they are normally applied. Or, a writer may take the structure of a psychological term—the ways people use it in everyday situations—and map that grammar onto brain structures, in a kind of literal spatial analogy. We will also have a look at other interesting devices used in this general approach.

In the second main approach, taken up in chapter 4, the writer recognizes something problematic in the neuroistic endeavor, and actively wrestles with the separateness of the two discourses. The effort to deal with the actual, complex uses of everyday psychological terms such as emotion or memory may lead the writer to split up the term and replace it with several new entities. He or she may redefine the psychological term wholesale, or rule out some aspects of the term's everyday use. In such ways, the neuroist writer tries to make the psychological term compliant with neurological language. Having redesigned the term's native grammar, however, the author typically continues to use the term in its usual ways, thereby circling around the central problem.

Instead of denying or wrestling with the confusion, some writers embrace it. In this third approach, explored in chapter 5, mystery may be highlighted as if it were a positive virtue. Ideological rhetoric—for example, about man's essence or spiritual nature—may be intensified. This is a way of trying to immunize the neuroistic narrative against critical analysis.

As will become apparent in these chapters, neuroist writers come from a variety of professional backgrounds. It is irrelevant whether the writer is a neuroscientist, a professional psychologist, or a journalist. The basic devices are the same. Taking the plunge into this literature means taking up specific pieces of writing, but the purpose is not to criticize any individual writer. Rather, it is to look at numerous instances of failure in order to show that no neuroist story holds up.

# 3

## Problem? What Problem?

We begin with accounts that glue everyday mental language together with neural language as if doing so were just a straightforward matter. These accounts do not close, but instead casually ignore, the gap between psychology and neurobiology.

### WORDS OUT OF CONTEXT

Some phrases and words can have both neurobiological and non-neurobiological meanings. Ignoring the separateness of the two contexts makes the distinction between neuroscientific and psychological meanings seem to vanish.

Consider, for example, an account of how children deal with distress by anticipating relief from a caregiver (Schore 1994). The author of the account, a psychologist, argues that the orbitofrontal cortices allow the child to calm himself in anticipation of relief. He supports this by quoting a phrase from the neuroscientist Joaquin Fuster to the effect that the prefrontal cortices "bridg[e the] temporal gaps" (1994, 344).[1]

The context of Fuster's remark, however, is the laboratory. The phrase is a condensed description of sustained neural activity subserving short-term memory in monkeys, during a delay period between a sensory cue and a response. When the psychologist claims that a distressed child's expectation of relief allows the child to "bridge the gap," the "bridging" is different: it has to do with a person's being able to wait for something. The writer has glided over the vast difference in contexts. One phrase comes from our

14

everyday psychological accounts of being able to wait for something, while the other comes from a quite specialized kind of account, about neural activity in a carefully constructed laboratory experiment. The similarity between the bridged gaps is only due to a superficial similarity in the words, but the author casually uses it as evidence for the identity of psychological and neurological events.

The same device appears when this writer discusses the toddler's capacity for self-control. He claims that the orbitofrontal cortex is a system governing internal inhibition and he surrounds the phrase "internal inhibition" with multiple citations of the neuroscience literature. The implication is that self-control in our everyday language and internal inhibition in a neuroscience context are the same thing. But they are not. In neuroscience, inhibition pertains to measurable changes in neural activity; in everyday language, self-control is something we attribute to persons. Although it is made to appear that the neurobiological and the psychological concepts are the same, their correspondence is spurious. Ignoring the contexts makes it look as though there is no hard work involved in equating the brain and the mind.

Along these lines, it is common for the distinction between two different meanings of the word "unconscious" to be ignored. On the one hand, it is derived from the Freudian theory of the mind. On the other, it simply means "out of awareness." By obliterating the distinction between these connotations, some authors make a connection between the unconscious (Freudian) mind and the body, based on the idea that brain and bodily processes take place out of awareness.

But this unawareness is simply inherent in the way we normally speak of bodies, since they are things and not persons. Speaking of bodies in this way, as we do, has nothing to do with Freud's innovative suggestion regarding the mind. Nevertheless, a writer who hybridizes the two uses of the word "unconscious" in a term like "biological unconscious" creates the illusion of explaining something about the mind and body simultaneously:

> The bridge between the neuroscience of emotion and psychoanalysis is that both centre on unconscious mechanisms. Neuroscience asserts that emotion is processed independently of conscious awareness; not in the dynamic unconscious of Freud but in a biological unconscious governed by the rules and constraints of neural circuitry

and neurophysiology. Like psychoanalysis, neuroscience asserts that conscious feelings are but the tip of the iceberg. Truly meaningful information is often under the surface. For neuroscience, the physiological, behavioural and technological (PET scan, MRI etc.) findings are the manifest content of unconscious brain circuitry. (Pally 1998a, 349)

In this passage the author asserts a parallelism between the mind and the brain based on the phrase "biological unconscious." The phrase "under the surface" evokes Freudian imagery and also seems to allude to that which is physically out of sight—under the skull—until exposed by imaging techniques. (We will see later that the double use of the term "unconscious" occurs in other psychoanalytic writing as well, where it is taken up more self-consciously as a problem.) By quietly ignoring the separate contexts of the term "unconscious," the author here makes a false connection between psychoanalysis and neuroscience.

The sociologist Sal Restivo has shown that writers proclaim parallelisms by selectively ignoring aspects of certain fields. Then they assert that the two fields just express the same basic knowledge in different languages. Restivo notes that claims of parallelism illustrate "the dangers of generalizing ideas or concepts which (a) have not been rigorously defined within their original realm of application; (b) are, insofar as they are explicitly defined or rigorously conceptualized, specific to the substance, logic, methods, and theories of their original realm" (1985, 9). Lack of rigorous definition besets many neurobiological terms—not to mention psychological concepts such as "emotion" and "unconscious." Rigorously defined or not, mental and neural terms have specific contexts that can't be ignored.

The matter-of-fact, nonchalant approach to the gap between mind and brain can be taken a step further. An author may bypass similarities between neural and psychological language and declare links between the brain and mind without offering *any* justification, even a contrived one. For example, Schore writes of the "establishment of this more complex orbitofrontal internal working model [which] allows for its function as a more effective strategy of affect regulation via its involvement in a reciprocally coupled mechanism of autonomic control . . ." (1994, 344). The languages of psychology (internal working model) and neurobiology (autonomic control) are casually and comfortably linked here as though it were already established that the physical processes—mechanisms of autonomic control—were equivalent to a psychological entity.

To be successful, a mind-brain account has to prove that kind of equivalence, not assume it. The truth is, we have no idea how *any* brain structure is involved in *any* representation—not even in the neurophysiology of vision, where there are still only candidate theories for the problem of coherent perception. As Maddox points out in his recent survey of neuroscience, "How these different (sensory) representations are constructed in the first place and how they are used as a basis for decisions remains profoundly unclear" (1998, 280). If, after decades of intensive research effort, it is unclear how the brain gives us visual experience, then how the brain creates the internal working models beloved of psychoanalysts is downright opaque. But the author seems to take the equivalence of the physical account and the mental account of internal working models as already somehow established.

## INCONSPICUOUS ACCOMPLISHMENTS

There is another technique for creating the impression that there's really no gap between the mind and the brain. It is to translate the language of the mind into the language of the body. From there it is a short step to the brain, since body and brain are both physical things, physically connected.

In *Descartes' Error* (1994), Antonio Damasio makes a sophisticated argument for the role of the body in states of subjective feeling. The argument amounts to substituting the grammar of bodily actions for the grammar of mind. A key element of the substitution is the thesis that bodily states ("somatic markers") are responsible for both *feelings* and *covert tendencies to behave in certain ways*. By referring here and there throughout the book to dispositions and actions, Damasio quietly makes language about behavior (overt bodily activity) equivalent to language about perceiving (a mental event):

> Prefrontal networks establish *dispositional* representations for certain combinations of things and events.
> Perceiving is as much about *acting* on the environment as it is about receiving signals from it. . . .

> Signals arising in [images formed in early sensory cortices] are re-
> layed to several subcortical nuclei and multiple cortical regions; . . .
> those nuclei and cortical regions contain *dispositions for response* to
> certain classes of signals. (1994, 181, 225, 240–41; italics added)

This behavioral language would appear to align Damasio with the philosopher Daniel Dennett (1991), who said that *qualia* (the philosophical term for the felt nature of experiences) are nothing more than "mechanically accomplished dispositions to react" (386). However, whereas Dennett concluded that qualia can simply be thrown out as philosophically superfluous, Damasio holds on tight to the language of subjective feeling. Despite the statements above, which were meant to show that bodily actions and feeling are really the same thing, he continues to use the languages of body and of feelings side by side, placing them in varying relations to one another just as we do in our everyday language. He sometimes does this in slightly novel ways: for example, he says feelings are "about" the body—a use which makes them like a kind of picture of it.

Continuing to speak of feelings is important, for Damasio wants to make a new proposal about them. Whereas we usually speak of feelings as being in opposition to reason (or nonemotional thought, also called "cognition"), Damasio introduces a narrative twist— namely, that feelings influence rational thought:

> [F]eelings are first and foremost about the body; . . . they offer us
> the cognition of our visceral and musculoskeletal state. . . .
> [F]eelings have a say on how the rest of the brain and cognition go
> about their business. (1994, 159–60)

What Damasio accomplishes in *Descartes' Error* is to suggest an innovation in the way we normally relate feeling to cognition. He does not show that the grammar of feeling can be replaced with language appropriate to brains and bodies: although in some spots he inconspicuously suggests it, he reverts elsewhere to using "feeling" in its normal way. He quietly undoes what would have been a radical act—reducing the language of the mental to the language of the physical—by retaining the everyday mental term and simply suggesting modifications to the grammar of its everyday use.

Purporting to have made mental state language equivalent to the language of the physical ultimately lands Damasio in a logical contradiction. He defines emotion as simply equivalent to certain body states. He then has to make a distinction between emotional body states and all the rest—otherwise there would be no justification for saying the term "emotion" refers to some distinct kind of body state. Other body states are therefore posited, and called "background states." And how are they defined? Background states, he writes, are more restricted in range and correspond to body states prevailing *between* emotions (150). Thus, his demonstration that emotion is grounded in body states is empty. However, the elegance of Damasio's style tends to conceal the flawed logic beneath it. Emptiness is the inevitable result of *any* attempt—elegant or otherwise—to dissolve the language of the mental into that of the physical while simultaneously retaining it.

## BRAIN ARCHITECTURE AND THE
## GRAMMATICAL ARCHITECTURE OF MIND

Another device of neuroism is to analogize the conceptual structure of an everyday psychological term to the physical structure of the brain. The complex everyday grammar of a mental term is discovered in the brain itself. The discovery, of course, depends on constructing appropriate narratives of brain architecture. Since neuroscience is accommodatingly plastic, having no central theory (see chapter 8), it is all too easy to find ways in which the brain's functional architecture might match a grammatical architecture.

For example, the widespread idea that socialization and rationality constrain our primitive natures is often recast as the dominance of the cortex over subcortical brain structures. This oversimplified, ideologically based idea about brain structure is then used, circularly, to prove the cultural idea. A colorful instance comes from a book by the journalist Daniel Goleman: "One way the prefrontal cortex acts as an efficient manager of emotion—weighing actions before acting—is by dampening the signals for activation sent out by the amygdala and other limbic centers—something like a parent who stops an impulsive child from grabbing and tells the child to ask properly (or wait) for what it wants instead" (1994, 26).

Many readers of that passage will be unaware that the functions of the prefrontal cortex aren't definitely known—certainly not as definitely as the phrase "dampening signals for activation sent out by the amygdala and other limbic centers" implies.[2] In Goleman's account, however, neurological uncertainty is entirely eclipsed by the ideology of emotion. The relation of the prefrontal cortex to limbic structures is made to look as though it exactly dovetails with, and thus provides a biological basis for, our cultural ideas about emotional impulses and their containment.

In another account (Wilson 1998), the architecture of the brain's evolutionarily older and newer parts has been made consonant with an ideology of human nature that melds genius with animal traits, passion with rationality, to give an "optimum balance of instinct and reason:"

> The old brain had been assembled . . . as a vehicle of instinct, and remained vital from one heartbeat to the next as new parts were added. The new brain had to be jury-rigged in steps within and around the old brain. . . . The result was human nature: genius animated with animal craftiness and emotion, combining the passion of politics and art with rationality, to create a new instrument of survival. (1998, 106)

A variety of stories can be told using this basic trick of transposing psychological-ideological architecture into brain architecture. The reason almost any story will work is that the brain has an incredibly complex structure, yet neuroscientists have only the most tentative ideas about how the various regions function. Thus, there is little constraint on plausible narratives concerning brain architecture.

In still another account, for example, Turner (2000, 23, 52) believes human beings chafe against social constraints, but are kept in bounds by emotions. He casts this thesis in neural terms when he writes that the "neuroanatomy that we still share with the African apes and our last common ancestor" favors weak interpersonal ties and autonomy. Laid on top of this wiring, he speculates, the brain developed additional wiring to enhance emotions and thus create social solidarity. The brain with its older and newer systems thus embodies our core dilemma: we are caught between wanting to avoid immersion in the group and wanting social bonds. (Whether all readers would identify this as their own core dilemma is questionable.) He goes on to assert that, as a further enhancement of the system for social bonds, moral codes arose from variations and combinations of emotions. "Moral codes . . . have a biological basis in the organization within and between the neocortex and subcortical limbic systems." There is no actual evidence for this statement, unfortunately.

In a very literal equation of brain architecture and psychology, Schore (1994) writes that an "anatomical locus . . . is equivalent to . . . the controlling structure which maintains constancy by delaying 'press for discharge' of aroused drives that has been described in the psychoanalytic literature." Once again we see the confounding of two distinct categories. A locus here means a physical part of the brain. A "controlling structure" is a psychological entity derived from classical psychoanalytic theory. Now, if a locus in the brain is equivalent to a structure of the mind, the mind-brain problem has been solved. But of course it hasn't been solved: the gap between the two areas of discourse has been quietly papered over by a superficial semantic similarity between "locus" and "structure."

## WHEN A NEW NARRATIVE LOOKS LIKE A SCIENTIFIC DISCOVERY

Neuroism reaches its most sophisticated form when proposals for *shifts* in the psychological grammar are offered together with neurobiological window dressing. The shift then looks like—but is not—a scientific discovery. An example is Damasio's proposal that emotion is reasonable and reason is emotional. He translated this psychological formula into the relations of various neural pathways, cortical and subcortical:

> At the beginning of this book I suggested that feelings are a powerful influence on reason, that the brain systems required by the former are enmeshed in those needed by the latter, and that such specific systems are interwoven with those which regulate the body.
> The facts I have presented generally support these hypotheses, but these are hypotheses nonetheless, offered in the hope that they may attract further investigation and be subject to revision as new findings appear. . . . It is as if we are possessed by a passion for reason, a drive that originates in the brain core, permeates other levels of the nervous system, and emerges as either feelings or nonconscious biases to guide decision making. (1994, 245)

To say that emotion is really rational, and rationality is really emotional creatively inverts the opposition these terms normally have: in our everyday language, "emotionality" and "rationality" are mutually counterposed and defined by their oppositeness. Damasio makes it appear as though this new, cooperative relation between them were an empirical rather than a grammatical matter by adducing neuroanatomical facts. He creates the impression of having made a scientific discovery.

But this story about reason and emotion is not compelled by the facts of neuroscience. In fact, it is the other way around. Neuroscientific facts are plastic and accommodating. Whenever the grammatical architecture of our psychological ideas is translated into a supposed functional brain architecture, we may be sure that the psychological account, and not the actual nature of the brain, is driving the narrative.

## SCIENCE MIMICRY

Another neuroist device which quietly erases the distinction between psychology and neuroscience is adopting the style of scientific articles. This can take the form of scientific-sounding language, abundant use of technical jargon, or illustrations typical of research articles.

For example, neuroist authors often use words such as "propose" and "postulate" when presenting their narrative ideas. These seem akin to the word "hypothesize"—but in science, a hypothesis is an idea framed in a way that makes it testable by experiment. Hypotheses suggest experimental tests, whereas neuroist proposals are offers to generate new narratives in which lay ideology is to be cloaked with selected bits and pieces of neuroscience. But "propose" and "postulate" have a scientific sound.

For example, d'Aquili and Newberg (1993) write, "We postulate that these [brain] areas, under certain conditions, may be involved in the genesis of various mystical states, the sense of the divine, and the subjective experience of God" (179). The "postulate" is that they can explain aspects of religious experience in terms of neuroscience. It's not a hypothesis to be confirmed or disconfirmed, but a proposal for a new neuroist narrative.

Another, somewhat crude, neuroistic device is to create the appearance of science through an abundance of scientific-sounding detail. As an example, the same authors present a very detailed neurological model of meditative states of consciousness, a model whose level of detail is entirely out of proportion to anything that is known empirically about how the brain functions during these experiences:

[W]e postulate that continuous fixation on the image presented by the right inferior temporal lobe begins to stimulate the right hippocampus; the right hippocampus, in turn, stimulates the right amygdala; the right amygdala stimulates the lateral portions of the hypothalamus, generating a mildly pleasant sensation. Impulses then pass back to the right amygdala and hippocampus, recruiting intensity as they go along. The impulses then feed back to the right prefrontal cortex. Progressively intense concentration upon the object reinforces the whole system. Thus, a reverberating loop is established, similar to that in the Via Negativa. (1993, 191, 193)

One is reminded of a Rube Goldberg device.

Similarly, discussing the heightened awareness of the meditator, a neurologist writes,

Impulses descending from the hypothalamus have only a few millimeters to go before they reach the midbrain. . . . Here are several circuits which might begin to blend a subject's heightened awareness into a large expanse of space. They include, for example, those nerve cells in the colliculi that have large receptive fields, and whose brisk responses are easily amplified to major proportions. . . . And just below the depths of the colliculi resides not only the long cylinder of central gray but also the two main sources of dopamine input. . . . Dopamine from the ventral tegmental area could play an obvious role in energizing the intensity and the effects of concentrative meditation techniques. . . . (Austin 1998, 506)

The writer expresses the hypothetical nature of his proposals honestly, with words like "might" and "could." However, the impact of anatomical figures in the book and detailed accounts of neurobiology such as this one work in the opposite direction, to instill a sense of conviction rather than tentativeness, of empirical science rather than speculation.

Some writers lard their texts with frequent citations of research literature. In neuroscience, research papers have many citations because they must refer to previous experiments and results. The presence of many names and dates in parentheses throughout a text gives a scientific appearance to the neuroist document—even though the content of the citations is often logically irrelevant to the narrative. (We shall see some examples of this irrelevance in chapter 6.)

Still another way that the aura of real, experimental science is created is through the use of pictures. Especially in recent years, impressive colored pictures of brain scans, both positron emission tomography (PET) scans and functional magnetic resonance imaging (fMRI) scans, have appeared not only in scientific research presentations, but in presentations made to lay audiences. Without knowing the real technical uncertainties, and ignorant of problems of interpretation due to experimental designs, laypersons are left with the simple impression that "seeing is believing."

In the book by Austin cited above, there is a color plate bearing a picture of a PET scan made of his brain while he lay in quiet awareness for two hours. He discusses the scan in an evenhanded way, together with evidence from scans done on persons doing yoga meditation which yielded different results. He discusses the limitations of the technology and the potential promise of fMRI for yielding more precise findings on changes during meditative states. Two different messages are being given. The text conveys tentativeness and uncertainty, while the pictured brain scans convey the concrete reality of an experimental demonstration.

When neuroist writers mimic the outward forms of research articles, the impression is of something captured and held in the hand: here is a diagram complete with arrows and boxes; here is a picture, richly colored, of a real brain; here are citations of research carried out in actual laboratories, underpinning detailed stories of how the brain works. But these are the trappings of science. Missing are hypotheses that can be tested, opportunities for others to replicate findings, and the real constraints that would be imposed if the neuroscience data related to a general theory of how the brain works—a general theory neuroscience doesn't have.

To summarize, we've reviewed some ways the basic logical problem of neuroism is kept out of sight. One device is to paper over the gap between psychology and neuroscience with words or phrases that sound the same but have different meanings within their own, proper contexts. Another is to proceed as though one can just translate the structure of psychological concepts into the physical structure of the brain. Finally, the narrative as a whole may be given the external features of a research article, creating the impression of a scientific demonstration of fact.

# 4

## Bringing out the Hammers and Saws

We've just seen that when writers casually piece mental and neuro-science languages together as though there were no real gap between them, the result is covert violence to context, logic, and the usual practices of science. Some writers, on the other hand, recognize the true difficulty of their task and openly use more vigorous strategies. None of their strategies succeed, but the problem is at least kept in the foreground for awhile.

### FIXING THE LANGUAGE: REDEFINING

One of these strategies is to wrestle with the multifarious, imprecise language of the mental, trying to tame it so that it becomes compatible with the language of neuroscience. This requires radically redefining the mental term in question. A good example is found in Joseph LeDoux's 1996 book, *The Emotional Brain: The Mysterious Underpinnings of Emotional Life*. LeDoux's efforts are strenuous, as they must be to deal with the complex natural grammar of emotion. They go as follows.

First, emotion in LeDoux's opinion is not unitary; one has to deal with individual emotions as separate entities. Second, emotion can be considered in its nonconscious aspects. Consciousness has to be added in order for there to be feelings, but he places the problem of consciousness outside the scope of his endeavor. Third, emotions are a set of special adaptive behaviors that are evolutionarily old and crucial to survival. Some emotions may be socially constructed but those are not under consideration. To understand an emotion, he says, it is better to consider universal behavioral functions than facial expressions or words because the former are less likely to be contaminated by social factors.

In sum, LeDoux defines emotion such that there is no general concept of it; and it is to be considered without recourse to either subjective experience or expressive behavior. This definition now bears little resemblance to what the everyday word connotes, for the everyday word's uses have primarily to do with subjectively felt states and their expressions. In fact, LeDoux explicitly acknowledges that he is avoiding the mind-brain problem by disregarding the subjective aspect of emotion. It would perhaps have been better, considering how altered this new concept is, to name it something other than "emotion."

To be sure, the focus on stereotyped, hard-wired adaptive behavior is entirely appropriate considered in the context of LeDoux's experimental work. In his key experiments, rats learn to expect electric shocks following acoustic tones. Once they have learned the connection, simply hearing the tones causes their fur to stand up and blood pressures to rise, among other signs of distress. In a series of very detailed and elegant experiments, LeDoux and his students have traced the brain pathways and chemicals responsible for linking the responses to the warning tones. Regardless of the high quality of the research, it's obvious from the nature of the experiments why ideas about subjective, conscious feelings or socially communicated expressions are irrelevant.

While LeDoux's radical redefinition of emotion massages it into compatibility with his research on fear conditioning in rats, that kind of fix really cannot be kept up—it is too much at odds with ordinary language. One can speak of emotion, or one can speak of rats' brains, but apparently one cannot speak of both using the same language. However, although LeDoux defines away the everyday connotations of the word "emotion" in his book, only retaining certain "adaptive behaviors," he continues to use the everyday word in his discussion. For example, on page 169, despite his previous statements that there is no unitary entity called "emotion," he is comfortably speaking of emotions and emotional disorders as if they could after all be addressed in a unitary fashion. This is the normal way of talking: indeed, a person interested in emotion might wonder why he or she had bought the book if the subject were not addressed in some recognizable way.

Now, if "emotion" starts off with a diffuse and complex grammar in everyday use, this is even more the case for "mind." Since the redefinition device we are considering picks and chooses among elements of a concept's grammar, mind is an even more fertile territory for redefinition than emotion. While some authors feel bound to address the problem of consciousness when mind is the issue, others are more idiosyncratic. One neuroist, for example, equates mind to the flow of information in the body: "I like to speculate that what the mind is is the flow of information as it moves among the cells, organs, and systems of the body. And since one of the qualities of information flow is that it can be unconscious, occurring below the level of awareness, we see it in operation at the autonomic, or involuntary, level of our physiology" (Pert 1997, 185).

This is not actually a speculation about the nature of something real but imperfectly known (that is, the mind); it is rather an offer to define mind in a particular way. The proffered definition hangs on the fact that we do not feel material processes going on in our bodies. Since material processes are in this sense unconscious, and there is an "unconscious" that is a feature of how we speak of mind, there is a superficial semantic connection there to be exploited. (We considered this neuroist device in chapter 3.) The author exploits this double meaning of "unconscious," using it to define mind in a new way: it's simply the body. The process is quicker, but no less radical, than LeDoux's selective redefinition of emotion.

How did the unconscious come to be a feature of mind? As is well known, Sigmund Freud added this new twist to the way people normally spoke of mind. Since Freud's term is used in various ways by neuroists in their arguments today, it will be worth taking a detour to consider what he actually did. As background, we should recall Wittgenstein's account of how we conceptualize mental life: we make pictures meant to capture, in a kind of shorthand, the complex ways we have of speaking of it.

## FREUD'S UNCONSCIOUS

Freud offered a *new* picture of the everyday language of the mental, just as a painter might come up with a novel method of depiction

that seems to show us something new about the reality of the de-
picted object. Freud's picture was a twist on the standard, and still
prevalent, picture of ourselves as somehow "looking inside" when
we consider the nature of first-person feeling. Just as Kant had
pointed out that our view of the outside world is not simply a true
and total reflection of it, but is conditioned and filtered by the properties of our sense organs, Freud said that our subjective experiences are like views of an inside world, of a "mental apparatus"
which is not directly knowable, but filtered in certain ways in order
to become the contents of consciousness.

The important point is that this *became a new picture* of our everyday grammar of the mind; it was a new "grammatical remark."
(Freud, however, thought of it as a scientific discovery.) It was a very
potent remark, for it doubled back on everyday language and practices to produce changes in the grammar of mind-related language
games. Thus, ever since Freud, the grammar of mind has included
the idea of the "unconscious"—and Freud's picture now haunts us
because it too is misapplied, such that we think there is something
real about the mind's unconscious that we must take up and explain.[1]

If Freud had dressed up his new account of the mind with bits of
neuroscience, it would have been a model for what neuroists do
today when they suggest innovations in the grammar of the mental.
(Early in his career, he did draw up a fascinating neuroist narrative—the "Project for a Scientific Psychology"—but he subsequently abandoned it, probably wisely.)

## FIXING THE LANGUAGE: SPLITTING UP

LeDoux's and Pert's approaches illustrate one way to force a psychological concept into compliance with neuroscientific research—
that is, by drastically redefining the psychological term. The noted
memory expert Endel Tulving (1995) has used a variant of this
"grammatical fix" technique: instead of redefining, he has broken
the psychological term down into other terms. In his introduction to
a suite of papers on memory, he begins by acknowledging that the
everyday word does not map onto the brain:

> Memory is a biological abstraction. There is no place in the brain that
> one could point at and say, Here is memory. There is no single activity,

or class of activities, of the organism that could be identified with the concept that the term denotes. There is no known molecular change that corresponds to memory, no known cellular activity that represents memory, no behavioral response of a living organism that *is* memory. Yet the term *memory* encompasses all these changes and activities.

Memory is a convenient chapter heading designating certain kinds of problems that scientists study. . . . (1995, 751)

So, although the everyday concept "memory" is used to organize a body of neuroscience findings, Tulving acknowledges it actually does not map onto any material process. This seems a pretty straightforward admission that there is a gap between the language game of the mental and neuroscience.

But since the job of the cognitive neuroscientist is to bridge this gap, Tulving has to supply a fix. The fix is to break the everyday grammar of memory down into more atomistic terms which may better capture what the brain actually does. He suggests that in the conglomerate, these neurally plausible subprocesses add up to something bearing a resemblance to our everyday term:

The quest for understanding of the identity and localization of the neural substrates of what we call "memory" has metamorphosed into the search for the neural correlates of encoding, consolidation, storage, and retrieval, separately for the different, dissociable forms of memory. (1995, 752)

However, even the neural basis of these proposed processes is not firmly established, as the following passage reveals:

Today's consensus holds that the limbic system, including the hippocampal structures, plays a critical role in certain forms of memory, even if that role and the exact identities of the "certain forms" are not yet clear. In addition the consensus holds that some other brain regions are also involved in memory processes. One of the most thoroughly studied of these other regions is the frontal lobes, although their role in memory has been somewhat controversial. (1995, 752)

Nevertheless, in the conclusion of Tulving's piece, a lid is lowered over this vat of ambiguity, in the form of optimistic statements about new techniques and the possibility of new insights into the basic character of memory. Like LeDoux, he manages to have it both ways, at first breaking up the grammar and then, in the end, keeping it:

> The painstaking early studies have now laid a solid foundation on which to build; the new techniques and methods will deepen our understanding of the workings of the brain/mind; and, perhaps most important, new insights into the basic character of memory, and the character of the scientific problem of memory, will help to guide research in even more rewarding directions. (1995, 753)

In the case of Tulving, LeDoux, and others, this two-step is necessitated by not acknowledging that everyday psychological language perhaps has *nothing at all* to do with the material things that science describes. Instead, by revising and dissecting the everyday terms, they suggest we will find a way to map emotion, or memory, or some other aspect of mentality, onto the brain. Radical revision of a term like "emotion" leaves us with something unrecognizable. Breaking a folk term like memory up into subprocesses results in equally intractable—or at least empirically questionable—terms, like "consolidation," "storage," and "retrieval." So the way forward using grammatical fix strategies is not at all clear, but never mind: we are assured that more research using better techniques will take care of residual ambiguities.

At its worst, the subprocess approach yields a kind of primitive phrenology. (Phrenology was a popular pseudoscience in the last century which traced personality traits to bumps on the skull, presumed to reflect the enlargement of certain brain areas beneath.) For example, the sociologist Jonathan Turner (2000) first tried to localize fear, anger and happiness to limbic structures, but as he couldn't accomplish this successfully, he reverted to chemicals: "Happiness and its variants are somewhat more elusive emotions, not being clearly located. . . . The fourth primary emotion—sadness—seems to be the result of neurotransmitter and neuroactive peptide release, or as is often the case, the failure to release substances promoting variants of satisfaction and happiness" (2000, 70). Anatomical phrenology did not work, so peptide phrenology was attempted. But since there wasn't enough to go on there either, Turner explained his psychological category, sadness, by the *absence* of peptides. When a neuroist speculates on the relation between psychology and the brain, the number of possible explanations is virtually limitless.

## FIXING THE LANGUAGE: NARROWING DOWN

Another grammatical strategy Le Doux has used on several occasions to deal with the problematic aspects of emotion is to draw a conceptual boundary line around central cases, excluding by fiat the problematic, ambiguous areas of the emotion concept. In 1990 he showed that a proposed brain system for emotion, a system that classically had been termed the "limbic system," did not stand up to scrutiny: he retained one brain structure, the amygdala, as being central to emotion, and declared that the contributions of the others were "unclear." Similarly, noting in the 1996 book that the set of basic emotions is difficult to define, he proposed concentrating on a few of the core cases agreed on by all, and leaving the "fuzzy" ones aside. Finally, he said there is no such thing as emotion as a unitary entity; rather, there are different adaptive behavior systems, each of which will have certain inputs, mechanisms for appraisal, and outputs. The one he had studied in detail, fear conditioning in the rat, was declared to be paradigmatic. The others, which involved brain regions other than the amygdala, were considered to be contaminated by various "nonemotional" factors having to do with the experimental setups.

What is actually likely to be the case—namely, that there are many, diverse behavioral survival systems involving many different brain structures—is eliminated from consideration by such logic. These systems remain outside the line drawn around the "clean" central examples that lend themselves to unambiguous localization.

Isolating central cases preserves concepts that have been stripped of their actual diffuse, ambiguous grammars. What remains is an artificially purified set of neural structures, a restricted set of basic emotions, and a definition of emotional behavior that hardly distinguishes it from behavior in general. LeDoux's strategy sanctions only one structure, the amygdala; one behavioral system, defense; and perhaps one or two basic emotions. This version of emotion may be scientifically docile, but it is a poor shadow of the majestic entity of our everyday language. The grammatical operation was a success, but the patient died. This outcome tells us that emotion really only exists through its use in our lay language and culture.[2]

And indeed, as we saw, LeDoux continues to use the everyday word throughout the book, reverting to the normal grammar of the concept in order to communicate with his readers.

## NAILING SEPARATE GRAMMARS TOGETHER

One of the subtler attempts at grammatical remodelling appears in the writing of the psychoanalyst Mark Solms (1997). Aware of the separateness of the grammars of Freud's unconscious and the grammar of physical processes, Solms labors to show that the two are meaningfully connected. His claim is that brain imaging and psychoanalysis study the same object, the mental apparatus and its functions, which are unconscious in themselves.

Let us take a look at the logic. Solms's dictum that the brain is "unconscious in itself" could simply be a restatement of our normal ways of talking about material things—namely, that they do not feel—and as such would not be a contribution to a philosophical discussion in which much ink has already been spilled. However, asserting a link between the brain and the mental apparatus as studied by Freud, through the useful contextual ambiguity of the word "unconscious," is not the conclusion, but the opening gambit of the argument.

According to Freudian theory, as Solms proceeds to remind us, an underlying mental apparatus is posited. We do not perceive its workings directly, but it nevertheless produces our subjective experiences. What Solms will suggest next is that since both the brain and the mental apparatus operate outside awareness, they must be the same thing.

Now, we saw in the comments on Freud above that the unconscious mental apparatus was *posited* through Freud's creative assertion that conscious processes are only what we are able to sense of unseen, underlying mental processes. The fact that the unconscious mental apparatus is an imaginative construct, while physical brains are not, is central to the huge gulf between their two language games. For, unlike Freud's concept, concepts about brains have links to the fact that we can weigh and measure them, hold them in our hands, reliably find them inside the skulls of humans and animals, and so on. Part of our concept about brains is that there are brains out there in the world whether we happen to be thinking of them at any given time or not—but our concept of feelings centers around their intrinsically subjective nature.

It is this latter, key grammatical difference that Solms attacks. He does so by using Freud's picture of consciousness: just as we can see an external thing like a brain, or a picture of one, says Solms, we have an *internal* sensory organ that produces qualia. He thus highlights the way the picture of external perception (in which the brain, among other things, can be an object) parallels the picture of subjective perception, in which Freud's mental apparatus is the object. Then, mistaking a similarity of pictures for a similarity of reality, and ignoring the profound differences in the actual language games pertaining to brains and consciousness, he declares the realities to be the same.[3]

In summary, Solms detects that there is a deep problem and proposes to solve it by cobbling two pictures together. Since neither the brain nor Freud's unconscious mental apparatus have phenomenal (subjective) properties, he says, they are really the same thing, and the mind-brain problem is solved.

Even if it worked, would Solms accept the implications of his own solution? If one takes the unconscious mental apparatus and the brain as one and the same, then the mental apparatus falls away as an unnecessary construct. Recent brain imaging experiments show how brain activity varies in precise accord with variations of conscious experience: given such results, it would seem that results from the brain itself can "take over from here" in explanations of consciousness (O'Craven and Kanwisher, 2000). There is no particular reason to believe that the underlying psychological processes Freud thought gave rise to conscious experience are there at all.

In fact, in due time, the current internalist concept of mind may be rejected. Brains will remain. But whether brains ultimately will be shown to have any of the properties we today mean by mental apparatus, cognition, or representation, is an open question, the answer to which could be "no."

## EUPHEMISMS FOR FAILURE: THE FINE PRINT DISCLAIMER

Finally, a device we can call the "fine print disclaimer" attempts to fix the fact that the neuroistic hammering and sawing isn't working. In this device, the neurobiological speculations are given at length and in detail; buried somewhere in the article, often near the end, is

a sotto voce statement to the effect that, of course, all this is speculation. The contrast between the salience of the speculations and the
inconspicuousness of the fine print is a bit reminiscent of sweepstakes mailings that declare in large letters "You are a winner!" but
in fine print tell you that, actually, you are not *yet* a winner. The "not
yet" analogue in neuroism is the prediction that as scientific research proceeds, speculation will give way to scientific certainty.[4]

In one example, the psychoanalysts Regina Pally and David Olds
(1998) write that certain theories about hemispheric function (the
idea of a right-hemisphere anomaly detector and of a left-
hemisphere interpreter) are speculative. However, they continue,
the theories "suggest a number of clinically relevant points," including that "What may occur during the interpretation of transference
is the engagement of the anomaly detector in the right hemisphere,
alerting the individual to the need to revise their neurotic belief
system" (987).

It is laudable that the authors first note that the underlying neuroscience ideas are themselves speculative; however, this does not deter
them from adding still another layer of speculation from the psychoanalytic arena. Presumably this piling on of speculation is acceptable
because the idea of an underlying neural basis for transference interpretation—a cornerstone of psychoanalytic practice—is so attractive
that its appeal eclipses the flimsiness of its speculative foundations.

In the concluding section of another paper in the same series,
Pally (1998b) writes,

> [N]euroscientists caution against over-zealous attempts to make spe-
> cific correlations between psychoanalytic theory and observation
> with neuroscience ones. We are not *yet* at the point where direct
> 'translations' can be made between the two fields. We need to learn
> what neuroscience has to offer, but we also need to understand that,
> at this point in time, neuroscientific explanations of analytic theory
> and treatment can only be provisional and tentative. (1998b, 575;
> italics in original)

Having delivered her caveat, Pally proceeds to offer yet more accounts of clinical phenomena in terms of left and right hemisphere
functions. Once again, the abundant details of the speculations
stand in contrast to the brief "fine print" disclaimers.

Another example of "fine print" comes from an article by the psychiatrist Judith Rapoport and the anthropologist Alan Fiske (1998). They offer detailed proposals for the brain circuits that underlie the symptoms of obsessive-compulsive disorder (OCD); they further suggest that OCD and religious ritual share a common neural basis. However, the authors confess that they cannot come up with a core or root cognition in OCD that is culture free, although they propose that, in general, most OCD behaviors have to do with boundaries, order, rules, and right or wrong. (But another authority on the disorder, psychiatrist Jeffrey Schwarz, characterizes the core cognitive problem as the pathological generation of error detection signals, giving rise to "a persistent internal sense that something is wrong" (1999, 113), which then produces repeated behaviors attempting to rectify it—an altogether different characterization.) Thus, Rapoport and Fiske have only a vague idea about the nature of the psychological-level entity that would link religious ritual on the one hand with brain function on the other. A sentence near the end of the article reads, "These highly speculative possibilities merit investigation because of their importance, if confirmed" (1998, 171). This is the sole reference to speculation within twelve pages containing considerable neural detail.[5]

There is one last fine print disclaimer that deserves mention as a neuroistic device. The weakest and most frequent kind of acknowledgment that the narrative does not really work logically is to say that the topic at hand is "complex." Like admissions of speculation buried at the end of an article, this acknowledgment often concludes a detailed, concrete neural account. Thus, Turner, at the end of a detail-packed chapter entitled "The Neurology of Human Emotions," writes, "The neurology of emotions is obviously more complex than the cursory outline presented in this chapter" (2000, 115). Such statements imply that if the neuroist effort is not working, it is only because the biology is so complex.

To summarize, in this chapter we examined energetic attempts to fuse the languages of brain and mind, attempts that bespoke an awareness of the fundamental gap between them. We saw that writers might redefine psychological terms wholesale, split them up into hopeful subprocesses, or prune them down to central cases, all in an effort to bring the language of the mental into conformity with the language and findings of experimental research. We also studied a clever attempt to equate the logic of psychological grammar (consciousness as a perceptual organ) with the grammar of biological description (physical perceptual processes). Finally, we saw that muted or open references to speculation and complexity indicated an ongoing, uneasy awareness of the basic problem.

In the next chapter, we shall see how unease is converted into awe. A writer who can't find any other way to deal with the logical gap can simply invite readers to dull their critical instincts by playing up an enjoyable sense of the mysterious, the paradoxical, or the magical. This is the third and final neuroist device to be considered.

# 5

## Let's Not Be Critical

We saw in the last chapter that a euphemism for the illogic of neuroist narrative is the word "complex." Illogic can be elevated to something more intriguing than complexity—for example, elusiveness. In a discussion of the design of the brain's circuits, Wilson (1998) repeats Damasio's narrative proposal that reason and emotion are intertwined: "Emotion is not just a perturbation of reason but a vital part of it. This chimeric quality of the mind is what makes it so elusive" (106). This hints at something that is tantalizingly hidden. Similarly, invoking the imagery of paradox and puzzle, Grotstein (1994), a prominent Los Angeles psychoanalyst, writes in the foreword to a neuroist work: "[The author] has highlighted how our neurons become key players in the formation of our personalities. We can almost now see brain and mind in a paradoxically discontinuously continuous Moebius strip connection" (xxii). Far from trying to cover over the confusion, the writer is piling on more, thereby inviting the readers to join in a frank celebration of mystery.

A passage by the outstanding neurochemist, Candace Pert (1997), extends the same invitation:

> The emotions are the connectors, flowing between individuals, moving among us as empathy, compassion, sorrow, and joy. I believe that the receptors on our cells even vibrate in response to extracorporeal peptide reaching, a phenomenon that is analogous to the strings of a resting violin responding when another violin's strings are played. We call this emotional resonance, and it is a scientific fact that we can

feel what others feel. The oneness of all life is based on this simple reality: Our molecules of emotions are all vibrating together." (1997,
312)

This writing is incantational. Words that evoke the sacred in
psychology (empathy, resonance, oneness) are packed together,
just as a liturgy may combine pictures of saints, sacred phrases, and
incense. The readers are beckoned to a sense of solidarity as par-
ticipants in these ritual elements—not least because the incantation
refers to solidarity in its very content, thus suggesting it.

## INVOKING FREUD

Another device that dulls critical instinct is reference to an icon of
authority. One of the main neuroistic uses of Freud, in this post-
Freudian age, seems to be to cast him as an early but handicapped
prophet, a genius who dimly glimpsed the Truth. Modern neuroscience, we are told, brings to light facts that were unavailable to
Freud—facts which now prove he was more right than he knew.
Seeming confirmation of Freudian theory is a device often used to
buttress the narrative's plausibility.

The problem with recourse to Freud, however, is that his theories
ranged from the general notion that there are laws governing mental
life to a multitude of particular ideas such as identification, screen
memories, repression, sexual desires of the child directed at his or
her parents, the death instinct, psychosexual stages of development
(oral, anal, and genital), castration anxiety, and on and on. Because
of the prolific body of work he generated, Freud can be used as a resource for any particular idea a neuroist happens to be focusing on
at the moment while Freud's other notions are left aside—or he can
just be cited in a general way, leaving it to the reader to decide which
theories are meant:

> If we accept the idea that peptides and other informational sub-
> stances are the biochemicals of emotion, their distribution in the
> body's nerves has all kinds of significance, which Sigmund Freud,
> were he alive today, would gleefully point out as the molecular con-
> firmation of his theories. (Pert 1997, 141)

[I]f a common circuitry were seen with that of OCD [obsessive-compulsive disorder], neurophysiology will have borne out Freud's speculation that "obsessional neurosis presents a travesty of private religion." (Rapoport and Fiske 1998, 170)

Freud surely deserves credit for pointing out the pivotal role of psychological defenses in helping us organize our mental life. Unfortunately, the theoretical schemes he constructed to explain them were nebulous and untestable. . . .

I've come to realize that [the patients] present us with a fantastic opportunity to test Freudian theories scientifically for the first time.

Indeed, we can pick up where Freud left off . . . and start experimenting on belief systems, consciousness, mind-body interaction and other hallmarks of human behavior. (Ramachandran and Blakeslee 1998, 3, 152–55)

At times neuroists suggest they are resuscitating Freud from his passé status. This aligns them with a readership which still—at heart—reveres Freud. Citing Freud is a device for instilling assent to the neuroist endeavor.

## MYSTERIES GALORE

Still another device is to hold up two mysteries at once, implying that they are connected:

The body is the unconscious mind! . . . The new work suggests there are almost infinite pathways for the conscious mind to access—and modify—the unconscious mind and the body. . . . (Pert 1977, 141)

The neuroist juxtaposes two mysteries here. One is the mystery of the mind-brain problem—the central mystery underlying neuroism—recast to include the body with the brain. The other is the mystery of the unconscious.

Because of its origins in Freudian theory, the unconscious retains faded resonances of the whole metapsychology—drives; repression; the ego, id, and superego—giving the word a kind of rich mystique. It connotes things about ourselves that we cannot know directly. In short, the word "unconscious" reeks of mystery. By holding up the unconscious and the mind-body problem at the same time as though they are connected, and indicating excitement with an exclamation point, Pert gives the impression that she has solved a problem or made a discovery—when in fact what she has done is to link two mysteries together. The same technique is used elsewhere in her work:

> I now know that this odd ritual was a powerful form of prayer, and I can only surmise that it acted through some form of "extracorporeal peptide reaching," a form of emotional resonance that happens when receptors are vibrating together in seemingly separate systems. This was before the term *subtle energy* had been used to describe a still-mysterious fifth force beyond the four conventional forces of physics. . . . (1977, 252)

Again, a connection is made between two mysteries—prayer and a special force in physics. The implication is that one mystery (the fifth force) can explain the other (the power of prayer). The narrative effect, as in the preceding example, is to provide a thrill for a mystery-loving audience. For this audience, conglomerations of mystery produce an even better experience than single mysteries. The device is like combining many sacred symbols at once to intensify a mystical experience. Like other devices discussed here, it drugs the critical senses, discouraging analysis of the narrative's scientific pretensions and logical flaws.

## EXALTING PARADOX

Now, mystery is of course unscientific. If present at the scientific dinner table, it is only as an unofficial guest that speaks in whispers, although the whispers may form an extremely seductive undercurrent of the conversation. A very interesting neuroist device involves relocating mystery, often in the form of paradox, to the chair of honor, thereby suggesting there is something positively virtuous about it. This device can be seen in a book about the neurobiology of mystical states of awareness (Austin 1998) written by a student of Zen Buddhism who is also a neurologist.

Austin states that he has tried to puzzle out his experiences of alternative states of consciousness brought on by meditation in light of his neuroscience background:

Indeed, one of my teachers recommended that I probe these intriguing experiences, using them as the focus for deeper questioning. Why was I surprised to hear this? After all, many Zen students incubate that other kind of riddle called a koan, and enter into a similar long drawn-out process of concentrated inquiry. (1998, xxiii)

This approach baldly rejects the usual scientific attitude, which is analytic and seeks explanation, in favor of a zenlike approach of nonexplanation, of immersion in paradox. Again, almost poetically:

Having now chosen to probe the complex interface between these two big subjects, we will be setting off to travel paths of incomprehension. The trip will take us along strange new planes that tilt away at improbable angles. (1998, 15)

Austin is one of the few neuroists to draw attention to a key practical problem for the neuroist endeavor—namely, the lack of a central theory for how the brain works. But, consistent with his rhetorical tack, he uses it to heighten the sense of the unknown:

Half-blindfolded and with mittens on, researchers are working to assemble a giant, shifting, three-dimensional jigsaw puzzle. Many of the pieces out on the table don't yet fit. So we can hardly expect that the two general lines of evidence—Zen and the brain—will always join in an orthodox way that satisfies most religious tests for authenticity and most formal scientific tests for proof.

As is so often the case, each new discovery in the neurosciences will only add more mysteries to the process. (1998, 5, 7)

Austin is using the double-mystery device as well; both Zen and brain are mysteries: "I invite the reader's caution: nothing about the brain, or Zen, is ever as simple as this book might suggest" (1998, preface).

As noted by Restivo in the case of those who attempt to draw parallels between mysticism and physics, "Parallelists stress the profundity, purity, and perfection of mystical knowledge. In effect, they immunize mysticism against the methods and theories of critical inquiry" (1985, 55). Having immunized both mysticism and the brain narratives against critical inquiry, Austin enters into an abundantly detailed neuroist endeavor, with many chapters of hypotheses about brain function and altered states of consciousness cloaked with pieces of neuroscience.

## HUMAN MARVELS

The spectrum of neurological disorders includes everything from concrete breakdowns of limb movement to strange alterations of thought and personality. On the one hand, a brain illness can impair fluency of motion or speech, or provoke abnormal movements such as tremors or sensations such as dizziness. We know enough about the geography and mechanism of many motor and sensory functions to make sense of such symptoms. On the other hand, because we know so little about the brain's general workings, neurological syndromes involving what we call "thought" baffle us. Distortions of the sense of self; trouble starting, organizing or stopping one's own actions; and complex delusions are just a few of the phenomena that so far elude neural explanation.

The reason they elude us is simply that we don't yet know how brains produce organized behavior, especially at what we call "higher" levels. Puzzling disturbances of thought in persons with brain injuries illustrate this fact. However, some of these poorly understood symptoms have been turned into illustrations of another kind: they have become potent symbols of mystery. Neurological patients with unusual symptoms have been converted by skilled communicators into texts, which these communicators—who are also neurological specialists—translate to the public. Although purporting to illuminate the relation between brain and mind, the true appeal of these patient texts is the sense of mystery they convey (just as in the written texts we examined earlier in this chapter). Therefore, the doctors who exhibit these human curiosities play up a sense of the bizarre and wonderful in their accounts:

It was almost as though inside Mrs. Dodds there lurked another human being—a phantom within—who knows perfectly well that she's paralyzed, and her strange remark was an attempt to mask this knowledge. (Ramachandran and Blakeslee 1998, 139)

But what of the status of the first lost, then recovered, memory? Why the amnesia—and the explosive return? Why the total black-out and then the lurid flashbacks? What actually happened in this strange, half-neurological drama? All these questions remain a mystery to this day. (Sacks 1985, 157)

This genre is exemplified by Oliver Sacks's book, *The Man Who Mistook His Wife for a Hat and Other Clinical Tales,* and appears prominently in Vilayanur Ramachandran and Sandra Blakeslee's *Phantoms in the Brain: Probing the Mysteries of the Human Mind.* It is closely related to the circus sideshows popular at the turn of the century, in which wild men of Borneo, fat ladies, albinos, Siamese twins, Chinese giants, midget triplets, and armless and legless wonders were displayed. As explained by the sociologist Rosemarie Thomson,

> The ancient practice of exhibiting anomalous bodies in taverns and on streetcorners consolidated in the nineteenth century into institutions such as American circus sideshows or London's Bartholemew Fair, where showmen and monster-mongers proliferated in response to a seemingly insatiable desire to gawk contemplatively at these marvelous phenomena. (1996, 2)

The hallmarks of the displays of odd human bodies included learned lectures by a showman or "professor" who also managed the exhibited person; textual accounts in pamphlets or newspapers that dramatized the person's life; and elements of staging such as feathers, animal skins, or jewelry, all in a spatial relation to the audience that served to set the exhibited person apart (1996, 7).

Whereas the human curiosities seen in sideshows challenged conceptions of the normal human body, contemporary neurological cases challenge conceptions of the normal mind. Sacks's *The Man Who Mistook His Wife for a Hat* is in fact a sideshow of human curiosities of the psychological sort, displayed for the wondering amazement of readers. The devices of freak display are amply present in Sacks's narratives. For example, he recounts this conversation with a patient:

> "Why, I guess I'm nineteen, Doc. I'll be twenty next birthday." Looking at the grey-haired man before me, I had an impulse for which I have never forgiven myself—it was, or would have been, the height of cruelty if there had been any possibility of Jimmy's remembering it.
>
> "Here," I said, and thrust a mirror toward him. "Look in the mirror and tell me what you see. Is that a nineteen-year-old looking out from the mirror?"
>
> He suddenly turned ashen and gripped the sides of the chair. "Jesus Christ," he whispered. "Christ, what's going on? What's happened to

me? Is this a nightmare? Am I crazy? Is this a joke?"—and he became frantic, panicked. (Sacks 1985, 24)

The tone of the account is sensationalist. And just as the sideshows often juxtaposed the bizarre and the everyday—for example
by showing the Armless Man drinking tea, or the Bearded Lady
posed beside her husband in a wedding portrait—the conventional-
ity of the showcased individuals is used to highlight their outlandish
features in these modern accounts as well:

> None of these people is "crazy"; sending them to psychiatrists would
> be a waste of time. Rather, each of them suffers from damage to a
> specific part of the brain that leads to bizarre but highly characteris-
> tic changes in behavior. They hear voices, feel missing limbs, see
> things that no one else does, deny the obvious and make wild, ex-
> traordinary claims about other people and the world we all live in. Yet
> for the most part they are lucid, rational and no more insane than
> you or I. (Ramachandran and Blakeslee 1998, 2)

To heighten the entertainment, the neurologist-showman accents
the horrifying aspects of the patients' mental lives:

> "Don't you understand? They mean nothing—nothing to me.
> *Nothing means anything* . . . at least to me."
> "And . . . this means nothing . . ." I hesitated, afraid to go on. "This
> meaninglessness . . . does *this* bother you? Does *this* mean anything
> to you?"
> "Nothing at all' " she said promptly, with a bright smile, in the
> tone of one who makes a joke, wins an argument, wins at poker. . . .
> In all these states—'funny' and often ingenious as they appear—
> the world is taken apart, undermined, reduced to anarchy and chaos.
> There ceases to be any 'centre' to the mind, though its formal intel-
> lectual powers may be perfectly preserved. The end point of such
> states is an unfathomable 'silliness', an abyss of superficiality, in
> which all is ungrounded and afloat and comes apart. Luria once
> spoke of the mind as reduced, in such states, to "mere Brownian
> movement." I share the sort of horror he clearly felt about them
> (though this incites, rather than impedes, their accurate description).
> (Sacks 1985, 112–14)

Sacks amplifies the bizarre into a sense of horror, simultaneously containing it within a setting staged to reassure us: *we* are the normal ones, the onlookers. Like the audience in a theater watching a horror movie, we can be both frightened by the imagined possibilities, and reassured by the fact that we are only spectators, participating temporarily until the lights go up again. Furthermore, professional experts have the situation in hand; doctors and scientists are at work providing rational explanations.

We said earlier that the subject of these patient texts is the mind-brain mystery. However, the true entertainment value of the accounts derives not from the sense of mystery but from the anxiety they provoke. The covert subject matter is the nature of the person. These stories both violate and reveal our collective concepts of personhood, our deeply held assumptions that persons, by definition, have intact autobiographical memory, knowledge about their own bodies, and sets of commitments to the meanings of things, for example. By breaking the rules of personhood, these anomalous cases challenge the assumption that personhood is a given, external reality, independent of our own construction. This is profoundly unsettling.

Thus, in the final analysis these stories confront us with the strangest reality of all—the *reality of our construction* of human life. The neurologist-authors ascribe the unwritten rules of personhood and their violation to the brain and its dysfunctions, obscuring their true origins in the realm of collective thought. Like the other neuroist texts surveyed so far, these clinical accounts concretize and naturalize psychology by placing it in the brain.

# 6

## Getting It Wrong

In the preceding three chapters we saw that neuroist strategies can't eliminate the fundamental gap between the languages of mind and neuroscience research. To review, the first basic neuroist approach to the gap was to act is if it weren't there. The second major strategy was to acknowledge the gap by strenuously remodelling the psychological grammar (but then reverting to the normal usage). The third was to abandon serious efforts to bridge the gap altogether and play up a sense of mystery instead.

We now turn from failed strategies to another kind of mistake found again and again in neuroist writing. The mistake is failing to interpret or understand the neuroscience literature correctly.

There is no uniquely correct way to synthesize the basic research literature in most areas of cognitive neuroscience, so there is considerable leeway for interpretation. This in itself is a problem. However, neuroists readily move from creative interpretation to playing fast and loose with the underlying experimental data in ways that simply are not obvious to readers without research backgrounds. In this chapter, we shift to lengthier, more close-grained analyses of neuroist writing, using a few representative articles to illustrate this kind of error and the difficulties involved in detecting it.

### NEUROSCIENCE'S SHIFTING SANDS

Because neuroscience is still in search of a central theory (as we will see in chapter 8), concepts appear and disappear as fashions change. When a neuroist dresses up a psychological topic with a currently fashionable neuroscience concept—such as reentrant activity or stable attractors—the narrative at least has a contemporary look. The transience of neuroscience fashions is highlighted, though, when a writer happens to adopt a concept that's become passé.

For an illustration, let's examine how Schore (1994) defends his proposal that the onset of self-control in toddlers is due to experience-dependent rewiring of the orbitofrontal cortex. He cites a 1965 paper by the neuroscientists Velasco and Lindsley to the effect that the orbitofrontal cortex is "a system governing internal inhibition" (1994). We might surmise from this that understanding internal inhibition as a neurobiological entity would give us a firm handle on the issue of toddlers' self-control. Turning to the Velasco and Lindsley paper in that expectant spirit, what do we find?

First, typical of research reports, we find a context that is technical and specialized. Velasco and Lindsley were reporting some unexpected effects of surgical destruction of the orbitofrontal cortex of cats. What they found was that cortical recruiting responses (negative electrical potentials measured by electroencephalography, normally elicited by midline thalamic electrical stimulation) were suppressed; and that spindle bursts (waves of cortical activity occurring regularly at a particular frequency, normally produced by reticular formation lesions) were also suppressed.

We also find that interpreting these data was not a clear-cut matter, even for the scientists themselves. To explain their findings, Velasco and Lindsley cited the neurophysiologist Horace Magoun (1964), who proposed that the orbital region was involved in inhibition. In the work they cited, Magoun adduced several lines of experimental evidence for his idea:

> A pronounced locomotor hyperactivity after frontal cortical lesions, referred by Ruch and Shenkin to ablation of orbito-frontal area 13, must obviously be attributed to removal of inhibition upon neural systems serving motility. In conditional reflex experiments, still another example of disinhibition following frontal lesions is provided by the repetition of trained movements hundreds of times between trials, during an experimental session. (Magoun 1964, 68–69)

Magoun then went on to cite evidence which was more equivocal, for example, that animals with frontal lobectomies were able to delay their responses in the dark but not when the test situation was brightly lit, invoking both external and internal inhibition by way of explanation. The line of argument ended inconclusively.

Magoun next related internal inhibition to sleep, which is the context in which Velasco and Lindsley refer to his work. Magoun wrote that large slow waves, recruiting responses, and spindle bursts in the electroencephalogram "characteristically bear a close relation to internal inhibition, behavioural drowsiness and sleep, although they can display dissociation from such behavior" (173). He had a larger picture in mind with respect to the concept of inhibition. It was based on the classic work of the neuroscientist Sir Charles Sherrington, whose influence peaked in the 1920s and 1930s:

> If the inferences drawn from these many contributions are correct, it is now possible to identify a thalamo-cortical mechanism for internal inhibition, capable of modifying activity of the brain partially or globally, so that its sensory, motor and higher functions become reduced and cease. The consequences of the action of this mechanism are the opposite of those of the ascending reticular activating system for internal excitation. The principle of reciprocal innervation proposed by Sherrington to account for spinal-reflex integration would appear relevant to the manner in which these two higher antagonistic neural mechanisms determine the alternating patterns of brain activity manifest as wakefulness and light sleep. (1964, 174)

What we found, in our quest for understanding how the neuroscience of internal inhibition might apply to self-control in toddlers, is a conceptual scheme derived from the study of reflexes, applied to hypothesized "higher antagonistic neural mechanisms," and finally brought to bear on the topic of sleep and wakefulness.

Not only does the actual research context of internal inhibition have no bearing on toddlers' self-control, the concept itself is no longer found in contemporary neuroscience—only in these older accounts. What has happened to the concepts of internal and external inhibition in the thirty years since Magoun's book was published—and for that matter, to the idea of opposing mechanisms for excitation (the reticular activating system) and inhibition (Magoun's proposed thalamo-cortical mechanism)? As global frameworks for understanding brain function, these schemes are simply no longer used, and they certainly are not being applied to the orbitofrontal cortex. Thirty years later, after much additional research, contemporary neuroscientists only claim, regarding the orbitofrontal cortex, that

[t]here are several indications that these regions are involved in the highest level of control of human behavior. . . . However, there has been little detailed information available on the anatomical or physiological organization of the OMPFC [orbital and medial parts of the prefrontal cortex], and the analysis of its function has necessarily been limited. (Price, Carmichael, and Drevets 1996, 523)

Had the author relied on this more recent, conservatively interpreted work instead of on Velasco and Lindsley's 1965 paper, he might not have found himself using an outdated concept derived from Sherrington's classic—one might almost say antique—theories of reflex. However, the phrase "internal inhibition" has a neurobiological meaning on the one hand, and sounds like a psychological idea on the other, which presumably made it attractive. (Elliding the separate languages of mind and brain on the basis of superficial similarity was one of the first neuroist devices we encountered in our survey.)

There is a treasure trove of concepts to draw on in constructing neurobiologically laden narratives. There are not only today's fashions, but in the secondhand shop of old neuroscience literature one finds concepts that can be turned to many a purpose. In the 1960s, "inhibition," as used by Magoun, already had a conceptual history and uses that traced back to Sherrington, as we saw. It was picked up by Velasco and Lindsley in an experimentally specific context; later, it was given still another slant in the story about toddler self-control. We see this concept, like a pawnshop guitar, passing through the hands of various owners, being creatively modified or even altered beyond recognition, relegated to the junk pile, rediscovered, and so on.

In neuroscience it's not only the concepts that are ambiguous: how various parts of the brain actually work is also uncertain. Many different things are proposed about every structure in the brain by the experts who are most familiar with the science involved. A non-neuroscientist will have little trouble finding a role for a particular structure that comports with the narrative he or she is developing.

For example, d'Aquili and Newburg (1993) quote the neuroscientist Karl Pribram to the effect that "prefrontal processes may function to produce redundancy in otherwise novel sensorial space" (183). This is different from the function Joaquin Fuster proposed (see chapter 3), which was to bridge temporal gaps. But d'Aquili and Newburg are developing a narrative in which the prefrontal cortex will mediate the occurrence of images during mystical experience, while the author who cited Fuster was telling a story about children's ability to wait.


## WHAT THE SOURCES ACTUALLY SAY

Sometimes the underlying literature is miscited. An example comes from the writing of d'Aquili and Newburg. In one part of their account, they draw heavily on a text by the neuropsychologist Rhawn Joseph (1990). They portray Joseph as proposing that the distinction between self and the external world is mediated by the left posterior superior parietal lobule. Specifically, the authors write,

> The fact that some neurons in the left PSPL respond most to stimuli within grasping distance and other neurons respond most to stimuli just beyond the arm's reach led Joseph to postulate that the distinction between self and world may ultimately arise from the left PSPL's ability to judge these two categories of distance. (181–83)

A perusal of Joseph's text, in the section entitled "Area 7 and the Superior-Posterior Parietal Lobule," (1990, 204–7) turns up the material to which d'Aquili and Newburg refer:

> In addition, many cells respond most to stimuli within grasping distance, whereas others respond most to stimuli just beyond arm's reach.

Joseph goes on to note how the neurons accomplish this:

> By integrating these [somesthetic and visual] signals these cells are able to monitor and mediate eye movement and visual fixation, map out the three-dimensional positions of various objects in visual space, and determine the relationship of these objects to the body and to other objects. (205–6)

However, one searches in vain here for the speculation attributed to Joseph by d'Aquili and Newburg.

Joseph continues, "Thus, the visual analysis performed by many of these cells is largely concerned with visual-spatial functions. . . . Conversely, when this area is damaged, depth perception, figure-ground analysis, and the ability to tract *(sic)* objects or to manipulate objects correctly in space (e.g., constructional and manipulospatial skills) are compromised" (206). There is no postulation regarding the distinction between self and world. A search further in the text, in the sections headed "Attention and Neglect," "Delusional Denial," and "Summary," where disorders of attending to extrapersonal space and of labelling parts of the body as belonging to self versus other are described, yields no postulate such as the one described by d'Aquili and Newburg. In fact, a neural basis for the ability to distinguish self and world appears nowhere in the chapter, nor is "self" to be found as an entry in the book's index. It seems safe to say, therefore, that the extrapolation from the data on neurons in the left posterior parietal region to discrimination between self and world is a postulate of d'Aquili and Newburg's, not Joseph's, who is more careful in his conclusions.

In fact, the basis in brain function of the distinction between self and other is entirely unknown. Nevertheless, if there are any speculations to be made about the neural basis for identifying self versus not-self, a neurologically informed writer would be more likely to consider right parietal functions, which are implicated in delusional denial of the left half of the body's being part of the self, as Joseph describes elsewhere in the chapter. However, d'Aquili and Newburg fastened on one result condensed (correctly) by Joseph from a number of basic neurophysiological studies, and made it the basis of a notion that the left posterior superior parietal lobule functions to distinguish self and world. They committed an error that neuroscientists try to avoid, not always successfully: while certain neurons may behave in certain ways, one cannot conclude that the "function of" that brain region is to carry out an analysis imagined in our own terms. (We don't know what the brain's terms are.) In this case, the behavior of certain neurons has been turned into a "probable" function—the self-other distinction—of central importance for the narrative d'Aquili and Newburg are about to spin.[1] Attributing their theory to Joseph was simply inaccurate.

Such errors are not likely to be detected by readers unable to track down the referenced work and read it with some understanding of the neurological and basic science issues involved. It is probably fair to say that most readers of the journal *Zygon,* where this article appeared, would be likely to take d'Aquili and Newburg's statements at face value. Indeed, they might feel some excitement at the "discovery" of the brain basis of the self versus not-self distinction.

## DON'T COPY SOMEONE ELSE'S ANSWERS:
## THEY MIGHT BE WRONG

Taking neuroscientists' syntheses as unproblematic can spell the downfall of even the most sophisticated theoreticians. For an example, we turn to the psychologist Paul Griffiths (1997) on the subject of emotion.

Griffiths understands the linguistic and social practices that sustain the concept. Regarding conceptual issues his sophistication is obvious:

> In everyday life, concepts are used to structure social systems, to further the interests of individuals and groups, etc. This suggests that the way in which the world is conceptualized by ordinary speakers will not simply conform itself to the most powerful explanatory and predictive taxonomy suggested by current science.
>
> [The concept of emotion] is utterly vague in the same way as concepts like "spirituality. . . ." It is relatively easy to imagine it falling into disuse. (1997, 7, 17)

Griffiths breaks up the everyday concept in order to salvage at least some of it. He writes:

> What we know about [emotional] phenomena suggests that there is no rich collection of generalizations about this range of phenomena that distinguishes them from other psychological phenomena. They do not constitute a single object of knowledge. Current knowledge suggests that the domain of emotion fractures into three parts. (1997, 14–15)

One part, he writes, are "affect programs," which are phylogenetically ancient, informationally encapsulated, reflexlike responses insensitive to culture. They are true neurobiological entities, unlike the other two, more culturally derived parts.

So neuroscience research will have to be used in this account. The problem is that Griffiths assumes empirical neuroscience is somehow cordoned off from contamination by lay concepts. However, the more we delve into the writings of Damasio and LeDoux, for example, the more we see that the "science" is penetrated through and through by everyday concepts and not cordoned off from them at all. Indeed, Griffiths' notes about Damasio's work reveal some awareness of this problem:

My worries about Damasio's proposal do not concern his fascinating neurological research, but the broad theoretical background that he uses his findings to support. As always, this framework is greatly underdetermined by the data. . . .[It] reflects certain background beliefs about emotion rather than anything in the evidence. (1997, 103)

In succeeding pages, Griffiths elaborates additional concerns about Damasio's conceptual overreliance on underlying, innate biological mechanisms.

However, despite his misgivings, Griffiths needs Damasio's authority to support his own ideas about innate affect programs. While Griffiths resorts to a summary in chapter 7 of *Descartes' Error*, his ambivalent reference to the limbic localization of the "phylogenetically ancient, pancultural elements of emotional response" as "fairly secure," based on that summary, suggests Griffiths is somewhat nervous about relying on Damasio's account.

He is correct in this, for there is as much interpretation in the parts of Damasio's short summary that Griffiths relies on, as in the parts that Griffiths worries about. Damasio (1994) writes that innate, preorganized emotions depend on the amygdala and anterior cingulate primarily; he lists Pribram, Weiskrantz, Aggleton, Passingham, and LeDoux in the context of animal studies—as well as Rolls, Davis, and Squire. Regarding studies of human patients, he lists Penfield, Gloor, and Halgren. He also refers to Klüver and Bucy "who showed that surgical resection of the part of the temporal lobe containing the amygdala created affective indifference" (133).

This single paragraph, which covers decades of research by many scientists, glosses over considerable issues of interpretation. First, the studies of nonhuman primates, including those by Klüver and Bucy, produced different results depending on whether the animal was caged or in social groups, a discovery originally made by the neuroscientist Arthur Kling and currently being elaborated in experiments conducted by the neuroanatomist David Amaral and his colleagues at the University of California at Davis. Second, in both nonhuman primates and human patients, there is an alternative to the classic idea of "affective indifference" as the explanation of the behavioral data following amygdala lesions. It is that the damaged amygdala, normally involved in interpreting the signals of others and generating bodily signals in response, leaves the animal or patient cut off from the flow of social signals. The classic account, which Damasio adopts in this highly condensed recapitulation, is not the only way to interpret the experimental data.

Another problem glossed over by this paragraph is that these various lines of research cannot be lumped together as if they were unitary. The experimental context of LeDoux's work is very different from the contexts used in the primate work: as we know, he studies rats' responses to learned associations between acoustic tones and electrical shock, a paradigm not used in monkeys and humans. As we also learned earlier, LeDoux has explicitly disavowed the idea that there is a general element called "emotion." In short, the underlying facts are much less unified and far more ambiguous than Damasio's account conveys.

Griffiths, in sum, despite his theoretical sophistication, makes the mistake of borrowing from an account which is already "packaged," to cloak his ideas with neurobiological validity. In subscribing to the idea that the "real truth" behind our language of the mental is to be found in neuroscientific investigation, he commits the core neuroist mistake. He compounds it by relying on a neuroscience source already permeated with interpretation.

Neuroist accounts are also used as sources by the journalist Daniel Goleman, author of the deservedly popular *Emotional Intelligence: Why It Can Matter More than IQ* (1994). The book's general theme concerns skills in reading and responding to social situations and how crucial these are to well-being. Chapter after chapter contain vignettes, many of them gripping, that highlight problems and remedies in the area of social communication. But are Goleman's statements correct with regard to the neuroscience?

He writes, "The amygdala's extensive web of neural connections allows it, during an emotional emergency, to capture and drive much of the rest of the brain—including the rational mind" (17). Goleman illustrates this idea with several stories about people committing fairly drastic acts before having fully assessed a situation, or having acted in ways that later turned out not to have been in their best interests. He terms this a "hijacking" of reason by emotion, and goes on to describe its neural basis, calling on LeDoux's research as a support. To see if this is valid, let's backtrack and revisit LeDoux's actual work once again.

To remind ourselves, LeDoux's research involves an experimental setup in which a tone is played to a rat in a cage, signalling an electric footshock to follow. When the rat learns the predictive nature of the tone, it shows typical defensive signs simply upon hearing the tone. This is called the "fear conditioning paradigm." LeDoux discovered that the conditioned response depends on a neural pathway which runs directly from the low-level auditory parts of the brain to the amygdala, bypassing the part of the cortex that is involved in auditory processing. By calling this an "emergency route," Goleman weaves a tale of how emotion can trump reason in an urgent situation. He is following LeDoux in this, just expressing it more dramatically: in his 1996 book, LeDoux endorsed the idea that emotional processing is more automatic and unconscious than what he terms "cognitive appraisal."

The point is that what both LeDoux and Goleman are communicating is all based on generalization from a specific experimental setting in which rats are exposed to tones and electrical shocks. The interpretive problem that gets glossed over is that this paradigm is probably not even applicable to *fear in general,* let alone to emotion in general. The footshock experiments are about rats learning to associate tones and shock. There is no guarantee that they speak to any more general mechanisms: LeDoux himself commented, in an article written for neuroscientists (not for a general audience) that many other ways of studying fear in the laboratory do not produce the same neural results as the tone-footshock experiments (1991). (To resolve this, he suggested that his paradigm might be the most pure measure of fear, but this seems somewhat arbitrary.) LeDoux carefully stated that emotion is not a unitary thing, that there are probably any number of specific mechanisms used to respond to specific situations. So by the same reasoning, fear may not be a unitary emotion.

And yet, there is a slippage, an almost irresistible pull to speak of emotion in general rather than restrict oneself to the specifics of the experimental setup. Part of that pull is exerted by features of the everyday concept that can't be found in specific experiments, such as the idea that emotions "make us do things" that are otherwise inexplicable and irrational; and that, despite this, they are the most noble part of our nature (as the opening vignette of Goleman's book, in which parents sacrifice their lives for their crippled child, suggests). It takes some heavy leaning on rats in cages to support these grand notions. But we are pulled nevertheless, because the rat experiments seem to prove what we already believe—namely, that emotion is a natural, biological part of ourselves. They pull our attention away from the fact that "emotion" simply designates a set of shared social practices.

Goleman's narratives, and all those used as examples so far in this chapter, do not succeed or fail depending on how scientifically accurate they are, for, as we are learning, accuracy is optional. What is important is adherence to the preexisting stock of conventional beliefs about emotion, and telling the audience in effect, "These are *really true;* neuroscience tells us so." Neuroscience does not tell us so, but the art of neuroism involves glossing over the troublesome experimental details, holes in the data, and logical problems.

## WHAT DO THE DATA MEAN?

On occasion, errors in neuroist narratives will be found in the design and interpretation of actual experiments. The psychologist Michael Persinger is a prolific neuroist in the area of mystical and religious experience. Not just a digester of neuroscience, he conducts psychological experiments himself, and interprets the results using neurobiology. Here, for example, are some results of an experiment regarding belief in past lives:

> Normal young men and women who believed they may have lived a previous life ($n = 21$) or who did not endorse ($n = 52$) this belief of "reincarnation" were exposed to partial sensory deprivation and received transcerebral stimulation by burst-firing magnetic fields over either the left or right hemisphere. Individuals who reported belief in reincarnation could be discriminated from nonbelievers by their more frequent report of experiences of tingling sensations, spinning, detachment of consciousness from the body, and intrusions of thoughts that were not attributed to the sense of self. (Persinger 1996, 1107)

We will leave aside very serious problems with the experimental design, and examine how Persinger interpreted his results. He makes a connection between imputed brain effects of the stimulation and the experiences reported:

> Individuals who endorsed an item indicating they believed they may have lived previous lives reported significantly more specific experiences while they were exposed to an experimental condition that

has been hypothesized to affect directly the neuronal networks at the magnitudes with which consciousness and the self are associated.

The significance of the clustering of these items [from the exit questionnaire, e.g., sense of a presence, dreamlike images] is amplified when one considers the cumulative literature regarding the consequences of direct electrical stimulation of the hippocampal-amygdaloid complex of temporal lobe epileptic patients. (1996, 1117)

One of the key features of Persinger's narratives, in this and other publications, is invoking one poorly understood phenomenon to explain another. For example, in the introduction to this paper, he cites his own previous work showing that students with exotic beliefs also had dichotic word-listening errors (dichotic listening is studied by presenting words or sounds to a subject through headphones, such that the right and left ears receive different stimuli). He suggests the errors are due to noise arising from structures of the deep temporal lobes and supports this assertion by citing a "moderate correlation" between the errors and reported experiences resembling those found in temporal lobe epilepsy. The experiences include sensations of smell, alterations of vision, feelings of déjà vu, and many other perceptual alterations. Such signs, and a "moderate" correlation, are very weak evidence from which to draw conclusions about the neural basis of dichotic listening errors.

More significant, the dichotic listeners—and the current believers in past lives—are nonepileptics, so the neural basis of these reports is purely a matter of conjecture. Persinger's demonstration may be completely circular, for the exotic beliefs and the subjective phenomena could both be functions of underlying personality styles that have no demonstrable basis in limbic pathology.

There are repetitions of some of the errors we saw earlier, in the form of misapplying basic neuroscience results to validate a psychological concept. Persinger's experimental procedure involves reducing auditory and visual input during the experiment, as we saw above. He says the rationale for reducing stimulation is to allow sensory neurons to be "recruited into other, transient neural networks which would amplify the effects of subtle (electromagnetic) induction to within the 'critical magnitude' that would allow the correlative phenomenon to merge or protrude within the awareness of the subject" (1996, 1111). He calls on a research article pertaining to epileptics to support this idea—but that article refers to the subjective correlates of actual, recorded epileptic discharges, not this proposed (but not demonstrated) induction of neurons into some vaguely defined networks. Reference to such an article after the sentence just quoted lends apparent credibility to the idea of neural activity producing subjective phenomena, even though the article is irrelevant to Persinger's ideas regarding networks and magnetic induction.

In all of these cases, the hard neuroscience used to support various claims about mystical experiences, emotion, or the basis of attachment and personality just isn't there. To show this, we've had to thread our way—often tediously—through the neuroscience literature, tracing claims and interpretations back to their sources. Most readers of neuroist literature are unlikely to undertake these tedious critical efforts, or even notice the need for them.

This concludes the critical survey portion of our look at neuroist writing. In the next chapter we take up the role neuroism plays in our culture, considering both its cultlike qualities and its alignment with commercial interests.

# 7

# Cults, Mysteries, and Money

## THE BRAIN CULT

We saw earlier that one of the neuroist strategies involves playing up the mysterious aspect of the mind-brain problem. The brain is both a piece of objective material stuff, and also—mysteriously—the source of human nature. The way neuroism describes the problem seems to be a secular version of the Christian mystery of the Spirit made Flesh. Far from being esoteric, however, neuroism is entirely mainstream, for the brain is in fact a sacred object in our contemporary secular culture.

What justifies the assertion that the brain is a "sacred object?" For one thing, we regard the brain with awe—not as just another organ, like the pancreas, but as an oracle of decipherable facts about human nature. When sounds made by air currents in underground caverns are understood as the voices of spirits, the spot whence they issue is treated with reverence, not just as a hole in the rock. The brain, like other sacred objects, is revered for its symbolic, not its merely physical qualities.

As with other oracular mutterings, the significance of data from neuroscience research laboratories requires special interpretation, by experts, to be made intelligible. There is a hierarchy of access to the inner workings of the mystery—the hallmark of a cult. Indeed, certain individuals can be considered the "high priests" of the cult, and it is from them that the authoritative ideology of neuroism issues. (Such hierarchies are found in other fields of experimental science as well.)

Access to a sacred object is a sign of status. In sociological language, neuroscience knowledge is "cultural capital." Its use can be understood along the lines described by the sociologist Randall Collins:

> [S]ome groups are more prestigeful, because of their accumulation of cultural and material resources for putting on impressive status displays. Hence individuals attempt to use their previous cultural capital to negotiate membership rituals in the highest ranking groups they can. . . . Not all individuals have the resources to enter all groups, and in some they may be pointedly excluded or subordinated. (1989, 20)

Nonneuroscientists attempt to negotiate membership within this high-status field by adopting neuroscientific rituals and trappings. They in turn serve as lower-order priests to the still less initiated. As we saw in the case of the "expert" on athletic performance, the appearance of being an expert on the brain is indeed a valuable cultural resource.

Violation of a sacred object is a moral matter provoking outrage. In the contemporary culture of neuroscience, it is close to contemptible to doubt that the neurons of the individual brain are the basis of all human phenomena.[1] "Real scientists" believe that human nature resides in the brain.[2] Dualism—the idea that there is material stuff on the one hand, such as the brain, and mind stuff on the other, such as spirit or soul—is anathema. There is a moral element to brain worship.

Why are we calling neuroism a "cult" and not a religion? Religions arise as cults mature: cult narratives are worked over by intellectual specialists, who devise a sophisticated system of thought. Early Christianity, for example, was a cult whose tenets were codified by theologians in the fourth and fifth centuries in the councils of Nicea, Constantinople, Ephesus, and Chalcedon in order to establish official doctrine.

At the present time, neuroism is a grassroots cult: more or less anyone is entitled to practice neuroism in his or her own creative way. However, the more neuroistic narratives come to be appreciated as valuable cultural resources, the more neuroism will invite attempts at control through doctrinization. Our society already has in place an apparatus for doctrinization, and that is our reliance on and submission to the class of "experts." As we shall see, an informal but powerful cadre of cognitive neuroscientists, increasingly influenced by the pharmaceutical industry, is shaping the doctrines of neuroism.

In sum, there are a number of reasons to think of the brain as a sacred object at the center of a cult. From the cult perspective, the mind-brain problem of philosophy is not a problem, but a mystery.

The rhetoric of this secular mystery is ubiquitous. We can find it, for example, in a university publication promoting research in one of its neuroscience laboratories: "As the physical site of our intellect, our emotions, our memories, and our dreams, the brain is the organ that defines our humanity" (*UCLA Medicine* 1998, 13). An awesome organ, indeed. The next sentence provides the requisite hint of mystery: "Its complexity baffles physicians. . . ." The article is about three-dimensional models of the brain: vivid color pictures appear alongside the text, demonstrating the impressive icons made by combining neuroanatomy and computer graphics.

Participating in the rites of the mind-brain mystery gives the feeling of participating in something of transcendent significance. This feeling can be put to use. The university's informational piece concludes by mentioning the donors who contributed to the research. In medieval times, wealthy patrons commissioned the painting of biblical scenes in cathedral chapels, and supported both artists and priests. Neuroscience laboratories and institutes are our secular chapels and cathedrals. Awe and mystery are good fund-raising tools.

The sociologist Emile Durkheim ([1912] 1995) showed that sacred objects acquire their transcendent qualities through collective but mundane human activity. As long as this activity remains outside our attention, it seems as though the object's grandeur is inherent in the object itself. To demystify the cult of the brain, we must turn to the activities of the people who study and interpret it.

## MAKING AND SPREADING THE CULT

Neuroscience researchers lay the groundwork for the mind-brain mystery through their construction of the brain as an objective bit of the natural world. They don't do so self-consciously, it is just in the nature of science as a practice. They also focus on specialized areas of research, as this is the best strategy for keeping their financial support. The brain, then, begins as a piece of objective nature, functionally atomized for the production of scientific careers.

If they have achieved high status within their own circles and happen to work on topics potentially of interest to the larger public, neuroscientists may find themselves in demand as experts. Experts are communicators. They are interviewed by lay media; publishers solicit them for books with mass-market potential; they command high fees as lecturers. Once thrust into it, this relatively select group of neuroscientists soon understands the opportunity niche: the more verbal and social ability an individual has, the more he or she can exploit it. With a good enough instinct for evocative symbols and issues in contemporary culture, a neuroscientist-communicator can appealingly intertwine the objective, natural science brain from the laboratory with a psychological story line. (As we've seen, the impossibility of achieving a real equation of the two languages can be converted by these communicators into experiences of mystery and awe for their audiences.)

This small, high-impact group has been able to appeal to general audiences, those educated readers who can read fairly long and demanding books but who may not have any training in science. Their works are also sources for the next tier of authors, who will bring the subject matter to an audience comfortable with less demanding reading—the popular market. Some books that belong to this second tier, such as *Emotional Intelligence* (1994) by the journalist Daniel Goleman, have been highly successful. The chain continues in stages into news magazines like *Time,* the Sunday feature sections of newspapers, and advertising that uses images of the brain or catchwords such as "right brain, left brain."

There is not a single, monolithic chain of discourse beginning with research work, transformed in the first tier by the premier communicators, then proceeding to ever more general audiences. Side chains develop in response to the needs of specialized audiences. In these chains, the writers may be members of the field for which they are writing; their accounts paste neuroscience onto the narratives particular to their fields. (As we've seen, examples in psychoanalysis include authors such as Pally; in sociology authors such as Turner or Fiske.) Like the neuroscientists and neuroscientist-communicators, such writers are responding to incentive structures, for wielding neuroscience concepts increases their status in their own fields. Examples of these audiences include but are not limited to psychoanalysts and other psychotherapists, managers, sociologists, anthropologists, and those interested in the interface between neuroscience and religion.

## Hidden commercial interests

Psychiatry and neuroscience, standing as they do at the interface of psychology and biology, can both be considered claimants to scientific authority on problems of human behavior. Psychiatry, however, has been weakened by its historic association with psychoanalysis: until a few decades ago, departments of psychiatry were dominated by psychoanalysts. In a huge shift, the leaders in psychiatry today are on the payrolls of the pharmaceutical industry. These individuals are senior teaching faculty, department chairmen, writers of standard reference works in psychiatry, authors of research studies on the effects of medication, officers of the American Psychiatric Association, authors of the standard diagnostic codes used by insurers—in short, opinion makers. But while the affiliations of the psychoanalysts were out in the open, the amount of support a lecturer, author, or teacher of psychiatry receives from pharmaceutical companies is between him or her, the company, and the Internal Revenue Service. There is a code of silence, not surprisingly, on the part of the individual recipients and the companies.[3]

The psychopharmacologist David Healy (1997), who is intimately acquainted with the shaping of the profession by pharmaceutical interests, has been able to show that current models of illness in psychiatry use concepts such as the "dopamine hypothesis" or "serotonin deficiency" not because these chemicals have been proven to be the basis of illness, but because they facilitate the marketing of drugs by pharmaceutical companies. One can think of such concepts as a specialized psychological language sculpted by commercial interests. This language has penetrated all the way into lay psychology: it is fairly common for people to ask psychiatrists whether their problems might be due to a "chemical imbalance."[4]

Healy writes, "In many respects the discovery of the antidepressants has been the invention of and marketing of depression. In the last decade, the pharmaceutical companies have struck out from the beachhead of depression into the heartlands of the neuroses, marketing obsessive-compulsive disorder, panic disorder, and social phobia as they have gone" (5). In psychiatry, we find a form of neuroism that is very obviously tailored to promote certain ideologies about what constitutes normal behavior, and to categorize abnormality in particular ways. There is an obvious money conduit—

although the amounts involved are a closely guarded secret—from industry interests to prestigous opinion makers who can articulate the ideologies.

It should come as no surprise that pharmaceutical interests are also finding ways to team up with cognitive neuroscience. This is already beginning to be seen in neuroist writing that cites the relevance of psychiatric disorders and treatment. For example, in a reference work on cognitive neuroscience, LeDoux attributes a growing research interest in the brain basis of emotion at least in part to "the fact that discoveries about the brain mechanisms of emotion have implications for understanding and treating the various emotional disorders that affect people" (Gazzaniga 1995, 1047). In the same work, the neuropharmacologist Floyd Bloom asserts that the presence of neurobiological correlates of "emotional diseases" in people justifies using animal models of emotion in research. While this logic is tough to grasp, it is clear in any case that both writers are linking cognitive neuroscience and the treatment of clinical conditions. (Bloom suggests clinical psychiatric disorders justify experimental neuroscientific approaches to emotion, but in the meantime, the diagnoses themselves are being justified by bits of basic neuroscience. This is logically circular, for the neuroscience of emotion and constructed clinical categories cannot justify one another at the same time.)[5]

It is obviously in the interests of the pharmaceutical companies that the diagnoses for which they are developing drugs should not appear to be "made up," but instead independently grounded in scientific facts about the brain. The following illustrates how this appearance is created, and how the medications are made to appear part of an independent scientific explanation.

A cover story in the *New York Times* Sunday magazine of 28 February 1999 begins with a vivid account of a woman suffering from anxiety symptoms, introduced to us as she is waiting to have a brain scan at a university hospital. We are told that she is being studied by the psychopharmacologist Jack Gorman, who is "using the latest high-tech imaging tools to study what fear, worry and obsession actually look like under the skull" so that he can study the effects of drug treatments on the brain.

Psychiatric researchers like Gorman are essentially at the clinical interface between the treatment industry on one side and laboratory studies involving animal models (such as those of LeDoux) on the

other. Gorman offers an account—which although entirely speculative, sounds definitive—of the neuroscience involved:

> "[Y]ou can think about how you tweak different parts of the system
> and you're going to get a different disorder," he says. "You know,
> bang on the amygdala and you're going to get panic attacks. Bang on
> the hippocampus and you're going to get post-traumatic stress disor-
> der. You mess up the medial prefrontal cortex and you're going to
> get too much worrying." (*New York Times* 1998)

Immediately following this quote the issue of a treatment rationale
is addressed. The writer of the piece euphemistically assesses the
issue as "complicated"—euphemistically, because the succeeding
paragraphs reveal there is no relationship at all. Instead, several
medications are mentioned, using their trade names rather than
their chemical names, as being "pretty effective for anxiety disorders." Gorman names two companies, Merck and Pfizer, who are
developing "a new generation of drugs" for depression and anxiety,
drugs that *may* work through effects on chemicals in the "fear circuitry." These vague statements are the sum total of the link
between the neuroscience and the drugs.

We see two elements here. One is the way in which speculation is
hidden in a mass of impressive clinical detail (a young woman law
student who could not drive across a bridge) and selective mechan-
istic accounts from animal models. The other, our focus here, is
how financial interests work to spin the web of ideology. The pharmaceutical companies need to promote a rationale for the use of
their products. To accomplish this, they heavily fund the work of researchers who on the one hand will promote their wares and on the
other provide a respectable looking scientific account both of the
disorders being targeted and the mechanisms of treatment.[6] Just as
in the neuroism we have explored in detail in cognitive neuroscience, clinical psychopharmacology is neuroistic—and the interests
at stake are obvious.

The neuroscience in this magazine article turns out to have been
window dressing: by the end, the scientific accounts have receded,
and the effectiveness of one particular medication, sertraline, is em-
phasized as the journalist returns to the details of the patient's biog-
raphy. No one actually knows how the medicine works, but its man-
ufacturer (Pfizer, which has provided support to Gorman) could

not have asked for a better public relations piece. It gives the impression, incorrectly, of having situated the illness and the medication within the experimental procedures of neuroscience. It is unfortunate that the journalist did not inquire about or comment on how much support the various researchers mentioned received from the companies and products that were highlighted in the article.

In sum, neuroscience narratives regarding emotion and clinical psychiatric narratives appear to be coevolving—which would be expected if a common market interest is penetrating both.

# 8

## An Unequal Match

In chapter 2 we learned why it does not make sense to describe a single grammatical remark (our everyday picture of the person) using two intrinsically distinct languages at once. Various neuroist attempts to get around the language gap—by ignoring it, wrestling with the psychological language, or frankly exalting mystery—were illustrated in chapters 3 through 5. Chapter 6 showed that nonneuroscientists frequently fall into more trivial errors as well, often through problems understanding the experimental neuroscience literature.

There is also a practical reason, and a serious one at that, why neuroism doesn't work. (We've referred to it in passing several times.) Experimental neuroscience is a warehouse of disparate facts about the brain, without a central theory of how the brain works. Away from the hoopla of enthusiasm for neuroscience, a few thinkers have been quietly remarking on this issue and its implications:

> The dirty secret of contemporary neuroscience is not mentioned . . . and is one I have not yet discussed. So far we do not have a unifying theoretical principle of neuroscience. In the way that we have an atomic theory of matter, a germ theory of disease, a genetic theory of inheritance, a tectonic plate theory of geology, a natural selection theory of evolution, a blood-pumping theory of the heart, and even a contraction theory of muscles, we do not in that sense have a theory of how the brain works. We know a lot of facts about what actually goes on in the brain, but we do not yet have a unifying theoretical account of how what goes on at the level of the neurobiology enables

the brain to do what it does by way of causing, structuring, and organizing our mental life. (Searle 1997, 198)

Yet the cruel truth is that the central objective of the now majestic
research program in neuroscience remains beyond reach: there is
only the most shaky understanding of how the brain, and the human
brain particularly, engenders mind—the capacity to reflect on past
events, to think and to imagine. (Maddox 1998, 276)

The annual American neuroscience meeting hosts around 25,000
scientists, and the amount of work presented there is commensurate
with those numbers. But despite the enormous data we are accumulating about the brain, broad theoretical suggestions are few and far
between. (Gold and Stoljar 1999, 827)

Facts only have meaning when they belong to an organized network. Otherwise, they are scraps of contextless information, like a
set of hieroglyphics for which there is no key. Without a basic theory of how the brain works, any given observation made in a laboratory floats free of an overall context.

At what is called the "natural history" stage of science, observations are simply collected, while the grand theory that unites and
explains them is deferred for later. This is a legitimate phase for
science. Neuroscience is in point of fact in the natural history
phase of its development. The vast majority of the huge number of
working neuroscientists, backed by billions of dollars of research
funds, are collecting natural history-type observations. This is almost certainly due, at least in part, to a reward system within neuroscience that makes narrowly defined experimental work more
sure of support.

To be sure, the recent abundance of functional imaging data has
stimulated some general proposals regarding how the brain works as
a whole. Cabeza and Nyberg (2000), in a review article, suggest that
the currently dominant "local approach," which relates particular
brain regions to particular cognitive operations, should be supple-
mented by "global" and "network" approaches. To sharpen the latter, theorists in neuroscience may have to call on concepts that have
been developed in other fields. For example, systems engineering
has produced theories of how a system is kept in a given state or on
a given trajectory; it also uses notions of hierarchy. In computer

science, systems architecture addresses issues such as which processes are active and what their relationships are; data types, data processes, and data encapsulation; and reliability. (The architecture of the most popular current model of brain function, neural nets, predetermines decisions about data and control without making them explicit.) It is certainly possible that the wealth of imaging information will compel cross-disciplinary theoretical efforts to understand the brain's overall functional architecture. However, enthusiasm about any particular set of concepts is always premature until considerable experimental work has conclusively shown that they are indeed embodied in brain function.

At present, neuroscientists are a bit like early astronomers studying the movements of heavenly bodies. Beginning as early as the fourth century B.C., astronomers saw that planets changed in brightness, that they sometimes seemed to stop and move backward and forward in a loop (this was termed "retrogression") and that their apparent velocities changed. The observed facts were complex and inexplicable without an equally complex theory. The Ptolemaic theory, which placed the Earth at the center of the universe, was therefore quite elaborate: each planet rode on the circumference of a small circle called an "epicycle," and the epicycles' centers revolved on large circles called "deferents." The Ptolemaic theory, which held sway for about fifteen hundred years, illustrates that there can be long periods in the development of scientific disciplines when observations are numerous, more are being accumulated, and a correct central scheme for ordering them is lacking.

Correct or not, however, there is almost always a theoretical scheme of some kind, which scientists continually try to improve upon, as Copernicus ultimately did successfully. What general theories scientists now have about how the brain works are even less developed than Ptolemy's understanding of the solar system; in place of a general theory are some notions, a bit like Ptolemy's deferents and epicycles, that we may call "part-concepts."

"Part" refers to the fact that a part-concept is grounded in specific, experimentally identified neural phenomena, but not in a general theory of brain function. Such phenomena can be patterns of neural firing under certain conditions, electroencephalographic recordings, quantities of neurotransmitter release, and so on. "Concept" refers to an idea that is more general than the phenomena themselves. Concepts are usually taken from a field outside neuro-science and may consist of mathematical ideas such as tensors, Fourier transforms, nonlinear dynamical systems, or information theory. They can also be simple, general ideas about neurons, such as the idea proposed by the neurophysiologist D. O. Hebb (1949) that neurons which fire in synchrony tend to become functionally connected; or the idea proposed by the neurophysiologist Wolf Singer (Engel et al. 1992) that information in groups of neurons is contained in their synchronized activity. Or they can be derived from more widespread, enduring cultural artifacts or ideas such as maps, searchlights, or hydraulic force. Sometimes concepts are based on technologies that are particularly captivating to society's imagination because they are new, such as telephones, computers, or holographic lasers.

The function of part-concepts is to organize neural phenomena and link them to psychological phenomena. "Channel capacity," derived from communication theory, was attributed to the brain in an effort to explain the auditory processing of speech (Broadbent 1958). "Homeostatic devices," derived from cybernetic theory, was applied to cortical neural activity to explain visual feature detection. The Hebbian hypothesis, among its many other uses, has been ap-plied to hippocampal neural activity in an effort to explain memory. The idea of synchronized neural firing has been applied to neurons of the visual system to explain the perception of objects. The con-cept of attractors, derived from a body of mathematical theory con-cerned with nonlinear dynamical systems, has been applied to the response properties of rabbit olfactory bulb neurons to explain ol-factory perception. Readers who are old enough will recognize that each of these part-concepts has had its heyday of popularity (Skarda and Freeman 1987).[1]

It is always possible that a given part-concept may turn out to hold the key to the overall function of the brain—but so far, this has not happened. A more typical fate is the discovery of the concept's limitations after an initial period of enthusiasm. Take, for example, the proposal for relating neural activity to mathematical concepts such as "chaos" and "attractors" referred to above. Several points emerged in critiques of this part-concept (Perkel 1987). One was the general point that mathematical models of various stripes have not held up because one eventually discovers that the mathematical framework has been dictating the biological assumptions. Another point was made by the mathematician René Thom (1987), himself a contributor to the mathematics involved. Thom equated the use of the chaos concept to "word play" and wrote that the term "has in itself very little explanatory power, as the invariants associated with the present theory . . . show little robustness in the presence of noise" (182–83). The commentators made the point that earlier descriptive images, such as Sherrington's evocative picture of a magic loom, were more or less equivalent to the concepts being wielded by the proponents of chaos theory.

That observation highlights an essential feature of part-concepts: they simply frame neural phenomena within *metaphors* that may or may not have anything to do with how the brain actually works. Because of their metaphorical nature, part-concepts are difficult to test in any definitive way. Many bits of laboratory data can count for them or be neutral, whereas it is difficult to find data that count against them. They are not rejected by being disproved, so much as dropped by the wayside as other, more attractive concepts come along. The research apparatus as a whole begins to look like a giant juggernaut that rolls along, throwing out data and interpretations which are never evaluated against any central theory. They sparkle like so much discarded tinfoil by the roadside, and are left behind as research fashions change. It is hard to tell if the juggernaut is moving in an intelligent direction—or just covering a space whose dimensions are unknown, in a random fashion.

## RIDING THE TIGER

Having considered the parade of part-concepts, we return to the lack of a central theory and its stark consequences. Normally, a field in its natural history stage does just fine collecting observations and trying to discover the laws that provide a framework for them. This is what it is supposed to do. But the presence of a vigorous, organized system of thought "next door" can threaten the whole process. Indeed, in contrast to the neurobiological part-concepts discussed above, psychological concepts are selected portions of a familiar, complex and well-established territory of ideas—the "language game" of mind.

If the organized system—psychology—is hungry for the legitimacy of an empirical science, it can offer its concepts to the fledgling science—neuroscience—in return for bits of objective data. Then what happens is this: When a new bit of neurobiological data is produced (for example, regarding what part of the brain is activated when certain types of words are shown to a subject) it goes by default into the conceptual framework of psychology, say, a theory of language processing or memory. This proceeding is very different from what happens in a science like physics, which has central theories such as the theory of relativity: there, experimental observations are used to count for or against the field's own theory. In neuroscience, data do not count for or against any model of brain function, for no model of brain function is being tested.

The union between psychology and neuroscience is not a marriage of equals. Rather than advancing its own separate agenda, neuroscience is used to deck out psychology's narratives. In principle, though, the imbalance between the level of articulation characterizing psychological ideas and the level characterizing neuroscientific ideas is temporary. If neuroscience *were* to develop its own central theory, and thereby some general power and coherence, it would cease to be merely a narrative resource for psychology, and would stand with it on equal ground. This would be a good thing. However, a meaningful bridge between the two would *still* be impossible, unless the fundamental problem of their two, separate languages could be addressed. (Chapter 9 presents an approach to this issue.)

By now the reader may concede that much of popular neuroist writing is a bit embarrassing. He or she may account for it by considering that it is done by nonneuroscientists, or by neuroscientists writing for nonneuroscientists. Perhaps when we are in the inner circles, closer to the research itself, we will find that neuroscience keeps psychology at a proper distance. But the truth is that researchers paste neuroscience facts onto psychological narratives just as much as popular neuroists do, only more subtly and soberly.

The philosophers Ian Gold and Daniel Stoljar (1999) showed this using Eric Kandel's very highly regarded basic research on the neural and synaptic bases of learning and memory. Using Kandel's own analysis of his results, they showed how psychological concepts were already intrinsic to the account:

> In modelling Aplysia neurophysiology . . . Kandel's theory appeals explicitly to psychological concepts. And it is this fact that leads us to

say that his model of conditioning is not solely neurobiological. The fact that Kandel's account is developed within an explicit and highly theoretical psychological framework means that the account does not provide a genuine alternative to psychological theory. It rather absorbs the required psychological ideas in order to provide a framework for understanding the behavior of Aplysia neurons and their role in conditioning. . . . [N]eurobiology has no concepts that can be used to describe the behavior of an animal so that the notion of a "pure" neurobiology actively in competition with psychology can only be a vision of some future science. (1999, 822)

This conclusion applies to all neuroscience accounts that reach toward explanations of behavior. Research articles written by and for cognitive neuroscientists are, as in the above case, permeated with psychological concepts. The authors also used the relation between long term potentiation (LTP), a property of synapses, and the psychological entity memory to illustrate this point. They demonstrated that LTP "can only provide the mechanism for a theoretical story that has already been articulated by psychology" (Stoljar and Gold 1998, 128). Even if LTP or a mechanism like it were to *replace* what we currently understand by the psychological process of memory in some scientific future, the process for which LTP is the mechanism would still have to be specified—and it would simply be some *new* concept of memory.

Interestingly though, Stoljar and Gold do not draw a pessimistic conclusion from their arguments. Their purpose is to show that a reduction to neural phenomena alone, without psychological-level descriptions, won't suffice to explain the mind. They think the two languages should inform and complement one another. They uphold the cause of cognitive neuroscience as an endeavor combining biology and psychology, in contrast to what they term biological neuroscience, which strives for an all-out reduction to the neural.

Stoljar and Gold write approvingly that cognitive neuroscience is "a science of minimal commitments; it holds that one ought to take what is known and make an attempt to integrate it with whatever else is known in an effort to develop a more adequate understanding of the mind" (1998, 130). The first problem, though, is that what is "known" in psychology, as we saw earlier, is known by virtue of entirely different procedures and in an entirely different way than what is known in neuroscience. The second is that psychology is as intricate and powerful as culture itself. A commitment to psychology is therefore not at all minimal.

For a picture of what the commitment to psychology actually entails, we can consider the following well-known limerick:

> There was a young lady from Niger
> Who smiled as she rode on a tiger.
> They returned from the ride
> With the lady inside,
> And a smile on the face of the tiger.

The lady in this verse, like Stoljar and Gold considering cognitive neuroscience, apparently wrongly assumed her ride would be a minimal commitment. Although their point is that psychology cannot be eliminated, these authors seem not to have appreciated how completely psychology swallows the mind-brain enterprise. From the moment the cognitive neuroscientist asks an experimental question, his or her approach to the brain is thoroughly permeated by psychological concepts. There is no reason to assume that these concepts have any relation to how the brain is actually organized and what it does. Thus, at the very kernel of the enterprise, the brain passively bears the imprint of all the psychological concepts the neuroscientist wittingly or unwittingly brings with him or her.

Here is an analogy. In the eighteenth and nineteenth centuries it seemed obvious that inanimate matter—mixtures of water and straw—could give rise spontaneously to living forms. In a series of ingenious demonstrations, however, Pasteur proved invisible air-borne organisms had seeded the mixtures. He showed that life was literally "in the air" rather than being a ubiquitous potential property of nonliving matter. Similarly, mind is "in the air" whenever scientists consider the material bases of human behavior. Today, cognitive neuroscientists unwittingly inoculate matter—the brain—with mind, and then find mind in the brain.

# 9

## Social Neuroscience

### PUTTING THE BRAIN IN PLACE

In the last chapter, we reviewed the pitfalls inherent in trying to join underdeveloped theories of brain function with robust ideologies of mind. Throughout the book we have focused on our mandatory participation in the social dimension and asserted that mind is created as a kind of social practice. So, are we pushing the material brain and body of the human being entirely to one side? How can we connect them with the mind? Or were the dualists, who spoke of mind stuff and brain stuff as two separate substances, correct?

What the dualists did not understand was the true relation between the mind and the brain. The key dimension of humanity, our participation in social forms of life—including the social practices that constitute "mind"—is the result of material processes that can be studied following the practices of natural science.

To show that this is so, we begin with the fact that persons conduct themselves in the medium of concrete, physical movements—those of the limbs, the head, the eyes, and the muscles of the face and throat. They wear or carry objects that have physical properties. Within the context of social forms of life, all these are components of acts expressing "person" in myriad ways.

Every such gesture evokes a finely tuned, equally physical response in others who witness it. Gestures are registered along sensory routes in the accumulating elaboration of the observer's physical responses, such as changes in heart rate, internally secreted chemicals, outwardly perceptible movement, or combinations of these. The body is a node in the dynamic web of social information.

Its exquisite sensitivity to social signals is starting to be appreciated as we learn more about how the central nervous system mediates the transformation of social input to output. Much of this information on the brain's social responsiveness is recent, and momentum is still gathering.

Before we survey what is known, a question arises: Why should the neurobiology of social cognition be taken more seriously than the neurobiology discussed in earlier pages—for example, the neurobiology of emotion, of mystical experience, or of memory? The argument that narratives from everyday psychology should not be naturalized by co-opting bits of an incomplete science of the brain would still seem to apply.

The response to the assertion that we may simply be naturalizing another arbitrary narrative boils down to arguments for relative legitimacy. True, we cannot escape from our collective stories about ourselves, including stories about our sociality, so we will continually run the risk of trying to naturalize those, whatever they may be. Nevertheless, it is an observable fact that we spend much of our time looking at one another's faces and bodies for signals, and listening to one another's voices; it is likely that we have conducted ourselves in this manner from very early on in the history of our species. Therefore, *this* activity appears more worthy of consideration for a neural basis than concepts about mind which may not even have been extant a century or two ago. Furthermore, whatever psychological concepts we now use had to have been generated through interaction: in other words, social interaction is logically foundational.

Even accepting that the exchange of social signals is both evolutionarily and logically fundamental, we still have to be cautious in drawing conclusions about what cognitive structures, if any, are determined by social nuts and bolts. For example, are there brain structures directly responsible for representing mental states—that is, for allowing us to have what is called a "theory of mind" about others? It may be more to the point to ask which structures are necessary, during development, for us to *learn* to participate in the set of social practices which refer to and enact having mental states of various kinds.

Finally, no matter how conservative or correct we are in our extrapolations from the building blocks of social interaction, we still do not escape from the incompleteness of neuroscience. That problem has to be solved hand in hand with the problem of formulating the right experimental questions.

Another argument for the legitimacy of a social approach to the brain is that its first empirical support came from investigations that asked entirely different questions. Indeed, initial observations of social brain activity were resisted for some time as being inconsistent with what neuroscientists believed about the brain. Tracing the course of neuroscientists' reactions to the first descriptions of "face cells"—discovered accidentally by visual neurophysiologists—reveals that they were brought to a reluctant acceptance of a brain specialization for face perception (Brothers 1997). It is a bit as though early geographers, beating about the bushes of an uncharted land expecting to find gryphons and cameleopards, stumbled across some actual creature that wasn't in the bestiary, a creature that defied received categories. Sociality thrust itself uninvited onto neuroscience: this history tends to lend weight to the argument that it is not just another folk psychological category in search of naturalization.

The remainder of this chapter is an overview of findings in the neurobiology of social cognition.

## EARLY DISCOVERIES AND PROPOSALS

In the 1930s, the neurologists Klüver and Bucy (1937) found that large surgical lesions in the front ends of the brain's temporal lobes produced quite unusual behavior in caged macaque monkeys. Among these was a change in the usual response to human handlers: whereas previously the animals had been quite aggressive, they now were docile. The social significance of these behavioral changes did not become clear until several decades later, however, when the zoologist and psychiatrist Arthur Kling performed similar experiments and observed his subjects in natural social groups rather than laboratory cages. The lesioned animals seemed normal in overall behavior, except for significant trouble responding appropriately to social cues. Based on these observations, and other studies involving related brain areas, Kling and his colleague H. Dieter Steklis (1976) proposed a specialized neural circuit underpinning the monkeys' capacities to form social ties.

Converging evidence for a social specialization came unexpectedly from the neurophysiological experiments referred to above.

Visual neurophysiologists were studying the responses of single neurons in the temporal cortex of monkeys to determine what features might trigger their activity. Entirely by accident, they found some neurons that fired selectively in response to the sights of hands and faces. A full decade elapsed before these findings were developed further, in part because they were controversial, in part because they lay outside the mainstream of vision research, and in part because the techniques for such experiments are extremely demanding and therefore not widely used.

During this period, psychologists suggested that the complex social environments in which primates evolved placed positive selection pressure on certain social abilities, especially the ability to deceive or manipulate (Humphrey 1983). In 1990, the "social brain hypothesis" developed these social and evolutionary themes somewhat differently. It held that significant "low level" stimuli, such as gaze direction and facial expression, are processed by dedicated neural circuits, and ultimately combined to yield a percept of person. Since mental states are an important part of the percept of person, the theory connected features of some clinical syndromes such as autism, in which an understanding of mental states is impaired, with then-emerging findings in primate neurophysiology. Brain structures in which it was proposed socially dedicated neural circuits would be found included the amygdala, cortical areas of the temporal lobes, and the medial cortices of the frontal lobes (all of which are interconnected.) A key feature of the theory was the idea that these processes are modular in the sense of being specialized for social stimuli (Brothers 1990).

The proposal placed special emphasis on the role of the face, noting the changes in facial expressiveness that have taken place during primate evolution. Comparative studies suggest that early primates were nocturnal, and relied on olfaction for social signalling. Their facial muscles were relatively immobile: for example, their upper lips were attached to the underlying bone as are those of dogs. As primates became more active during the daytime, and more visual, their faces became more suited for social signalling. The upper lip was freed, and the muscles around both the mouth and eyes were available for purely expressive, as opposed to instrumental, uses.

Of special significance for the experimental findings that will occupy us below is that the organization of the brain changed as well.

Its connectional architecture shifted so as to bring visual information to a structure that had previously specialized only in olfactory information, the amygdala. Comparative anatomists have found that there is an increase in the relative size of the amygdala along the lineage leading to our own species, in step with the huge expansion of cortex from which the amygdala receives inputs (Stephan, Frahm, and Baron 1987). These facts and proposals provide the context for a variety of "social brain" discoveries made in the last decade, to which we turn next.

## SOCIAL NEUROSCIENCE

In the remainder of this chapter we review reports of brain activity selectively responsive to the following stimulus categories: faces, voices, movements of the eyes and mouth, meaningful facial expressions, and expressive body movements. Moving from the processing of simple expressive elements to more complex percepts, we then review what is known of the brain's role in creating the idea of mental states.

Human functional magnetic resonance imaging (fMRI) consistently shows that an area of the ventral posterior temporal lobes adjoining the occipital lobes is specifically activated by the sight of faces. Concentrated around a patch of cortex on the bottom surface of the brain named the fusiform gyrus, it has been dubbed "the fusiform face area" by Nancy Kanwisher (Kanwisher, McDermott, and Chun 1997), based on her studies. Another group has confirmed the face-specific activity of the fusiform gyrus using recorded electrical potentials (Puce et al. 1997, Bentin et al. 1996). Using fMRI, these investigators showed that face-specific brain activity is also found in the neighboring inferior temporal gyrus and occipitotemporal sulcus (Puce et al. 1995).

How the brain processes the sounds of the human voice is much less understood than how it processes faces. However, the idea that certain areas of the brain are specialized for the sight of faces as opposed to other visual stimuli has recently found its counterpart in auditory studies. A region of cortex on the upper bank of the central superior temporal sulcus, on the lateral surface of the brain, has been found in an fMRI study to respond selectively to voice as opposed to nonvoice auditory stimuli (Belin et al. 2000).

Although static images of faces seem to be the best stimulus for the fusiform area, the sight of eye and mouth movements stimulates activity in regions of the temporal lobes adjacent to the "voice area." Using evoked potentials, Bentin and his colleagues (1996) inferred a neural source of sensitivity to the sight of human eyes separate from the source sensitive to faces. In other experiments using fMRI, they tested subjects with stimuli that consisted of moving check patterns, moving eyes, and moving mouths. In contrast to the moving patterns, the moving eyes and mouths stimulated activity bilaterally in a region of the posterior superior temporal sulcus.

This region would seem to be the site where perception of gestures emanating from the face—the most common being those movements of the mouth and eye regions which normally accompany speech—are integrated. Indeed, it has been known for some time that speech perception is influenced by the sight of mouth movements. A recent imaging study showed that the sight of silently spoken words (lipreading) stimulates neural activity in the lateral superior temporal sulcus and the cortex within the sulcus, which is also where auditory processing occurs. Apparently this "secondary auditory area" integrates the sounds of speech with visual information from the face (Bernstein et al. 2000).

Bentin proposed that the more ventral (bottom surface of the brain), fusiform-gyrus-centered region might be responsible for assigning identity to faces, whereas the lateral system involving the superior temporal cortex might be responsible for decoding gestures of the face. The proposal seemed to be supported by another study which used static images conveying information about identity and gaze direction. The authors of the latter study concluded that the face-responsive region in the superior temporal sulcus responds to "changeable" aspects of the face, whereas areas in the fusiform and neighboring inferior occipital gyri are responsive to the invariant aspects that underlie identity (Hoffman and Haxby 2000). However, a recent investigation has found that static pictures of faces with happy expressions provoke more intense activity in the fusiform gyrus than those with neutral or disgusted expressions, casting doubt on this dichotomy in its simple form (LaNoue et al. 2000).

All communication does not issue from the face; other movements besides those of the face appear to be processed in lateral temporal regions as well. Eva Bonda and her colleagues devised point-light displays of human motion. Their stimuli were based on the finding that a few light displays attached to joints of a moving body are readily interpreted by observers as showing the movements in question, even when darkness prevents any other features from being visible.[1] By using such displays, the experimenters were able to control for featural complexity across different stimuli. The stimulus set consisted of goal-directed hand action (imitating the act of reaching toward a glass, picking it up, and bringing it to the mouth), expressive whole body motion (dancelike movements), object motion, and random motion (Bonda et al. 1996).

Using positron emission tomography to measure brain metabolic activity, Bonda and her colleagues found that goal-directed hand action activated the left posterior superior temporal sulcus (an area slightly more posterior and superior than that sensitive to facial movement) and the cortex of the left intraparietal sulcus. This was consistent with previous formulations that the two regions form a system for producing and interpreting goal-directed actions in extrapersonal space. Bonda discovered a striking difference in the pattern of brain activation for the expressive stimulus, however. Expressive movements of the body produced activation of the amygdala and regions connected with it, especially in the right hemisphere—as well as the right superior temporal sulcus.

It is known from anatomical studies in monkeys that neurons in the superior temporal cortex and regions surrounding it project to the amygdala, a collection of nuclei in the anterior part of the temporal lobe.[2] As reviewed earlier, one of the first clues about the organization of social brain processes came from observing the social deficits of monkeys with amygdala lesions. We also saw that during primate evolution the amygdala shifted its connections away from a strict dedicaton to olfaction, growing in step both with the rest of the cortex, and the increasingly expressive faces in the social milieu. Given their history and connections, it is not surprising to find the amygdala is involved in processing some of the same social features as the superior temporal cortex.

Indeed, patients with injuries or developmental lesions of the amygdala are impaired in perceiving certain facial expressions, tones of voice, and gaze direction. Ralph Adolphs, Antonio Damasio, and their colleagues (1994) studied a patient with a rare congenital illness causing progressive bilateral destruction of the amygdala and found that she was impaired in recognizing certain facial expressions, especially fear. Another patient with bilateral destruction of the amygdala, consequent to surgical treatment for epilepsy, was found by Andrew Young and his colleagues (1995) to be similarly impaired in the discrimination of facial expression. This patient, furthermore, had difficulties interpreting eye gaze direction. Interestingly, upon further testing it was discovered that she was also deficient at understanding expressions conveyed by tone of voice, as well as prosodic (tone of voice) information distinguishing questions, statements, and exclamations (Scott et al. 1997).

More recently, patients with the clinical syndrome known as fragile X have been shown to have deficits in social behavior, especially in maintaining mutual eye contact. A group of patients was tested with pictures of faces whose eyes were averted or directed forward. Compared with control subjects, they had difficulty determining whether the gaze was directed at them. Subjects and controls also showed a difference in fMRI-detected patterns of brain activation: compared to controls, fragile X patients lacked activation of the amygdala, among other areas (Merin 2000).

Expressions of body, face, and voice, and the direction of gaze, when combined, are the building blocks of what we call "intentions." Intentions are mental states like "is thinking of," "is afraid of," "desires," and so on. We've just seen that the amygdala, with its "downstream" connectivity from more posterior temporal cortical areas, seems to be important when it comes to registering primitive expressive elements. So it is interesting to discover that the amygdala probably also plays a role in ascribing mental states.

The earliest evidence for the role of the amygdala in ascribing mental states to others came from observations made by the neurologist Pierre Gloor (1986). In the course of clinical tests designed to pinpoint the source of a patient's seizures, Gloor ran small amounts of electrical current through electrodes previously implanted in the patient's amygdala and nearby hippocampus. While such stimulation can produce a variety of physical sensations and feelings, what struck Gloor was the social nature of the feelings described by the patient. They could best be conceptualized as reactions stimulated by someone else's social intentions directed at him. The identity of the "someone" was vague, as might be expected in this setting: "Upon stimulating his left amygdala at 1 mA, he had a feeling 'as if I were not belonging here,' which he likened to being at a party and not being welcome. . . . Right hippocampal stimulation at 3 mA induced anxiety and guilt, '. . . like you are demanding to hand in a report that was due 2 weeks ago . . . as if I were guilty of some form of tardiness'" (1986, 164).

The artifically stimulated feelings can be thought of as the output of a system which sensitively matches social input—in this case, the intention of ostracizing him (as one would do to an unwanted guest) or taking him to task (as one would do to a subordinate derelict in some duty)—to the appropriate response (an urge to withdraw; an urge to apologize or make amends). Under everyday conditions, another person's indications that he or she is ostracizing or scolding would be based on multiple, lower-level social cues such as tone of voice, facial expression, and so forth. It is likely, given what we now know, that such information is initially processed and integrated "upstream" from the amygdala in temporal regions such as the fusiform face area and the superior temporal sulcus. Gloor's data make it appear that the amygdala is involved in a system that transforms the input into the appropriate bodily responses and their associated subjective feelings.

Although the language we use to describe these social intentions and corresponding responses is complex and culturally determined, it is not unreasonable to think their analogs—having to do with group acceptance in the first instance and dominance/submission in the second—are present in some form in all primate societies. Gloor's observations and their implications led me to coin the phrase "hot theory of mind" to distinguish such social intentions from the mental states normally invoked in theory of mind research, that is, states of belief or knowledge.

In the example above, the patient explained his feelings by reference to intentions directed at him by another. What about intentions directed by another toward a third person? It seems that the amygdala may be involved in this attribution too. The woman patient with the congenital disorder that destroyed her amygdala bilaterally, referred to above, was tested to see whether she attributed mental states to others in the normal fashion. The experimenter, Andrea Heberlein, showed her a short movie depicting geometric figures, such as triangles and circles, moving around in relation to one another. Asked to describe the movie's story, the patient said, "Let's see, the triangle and the circle went inside the rectangle, and then the other triangle went in, and then the triangle and the circle went out and took off, left one triangle there. And then the two parts of the rectangle made like an upside-down V, and that was it." Her narrative, in other words, was restricted to a physical description of the events. A normal subject, in contrast, said things like, "There was a large triangle chasing around a smaller triangle. . . . Finally he went in, got inside the box to go after the circle, and the circle was scared of him . . . and the big triangle got upset and started breaking the box open" (Adolphs 1999, 473). In other words, the normal subject "saw" mental states portrayed through the movements of the geometric figures.

Theory of mind means attributing states like hostility, fear, or amorous interest ("hot" intentions) to others, as well as states like attention, knowledge, and belief ("cold" intentions). Theory of mind might fail to develop, as in autism, because the basic circuitry for processing face and eye signals isn't initially organized in the normal fashion, leaving these children at a disadvantage when it comes time to acquire the socially taught conventions for assembling expressions and gestures into a lexicon of mental states. Alternatively, there could be a neural system that is responsible for perceiving automatically the mental states spelled out by behavior (in analogy to how we perceive the meaning of a word from the sight of its letters), a system of which the amygdala is a component. Damage to any part of such a system would cause a kind of blindness to mental states.

This brings us back to the nature of modularity. The modular social brain hypothesis was expressed in a slightly revised form in 1992 (Brothers and Ring). A careful, empirically based critique of its core claim for modularity of social processing was published subsequently by Annette Karmiloff-Smith and her colleagues in 1995. Their critique of what they term the Brothers-Ring hypothesis addresses the issue raised at the beginning of this chapter, namely, what higher-level social behavior shall we attribute directly to neural organization, as opposed to social learning?

We suggest that in normal development there are distinct, domain-specific, skeletal predispositions for discriminating stimuli relevant to language, face processing, and theory of mind. With the massive early experience of superimposed inputs (i.e., face, voice, and human interaction all take place in a superimposed fashion), these predispositions gradually take over privileged circuits in the brain that become increasingly specialized and progressively interconnected. They are not necessarily spatially adjacent in the brain, but both their macro-development and, subsequently, their on-line processing may be closely related in time. Such temporal co-occurrences can give rise, as a result of development, to the emergence of specialized interconnecting circuitry. As development proceeds in the normal case, a process of modularization . . . gives rise to the emergence of separate, modular-like organization for each sub-domain relevant to theory of mind. Subsequently, we speculate that the computations relevant to theory-of-mind representations in each domain may with time give rise to an emergent superordinate modular-like organization for the pragmatics of social interaction in general, along the lines of the Brothers-Ring hypothesis. (Karmiloff-Smith et al. 1995, 203)

A reasonable account of how theory of mind arises will consider the interaction between innate processing biases in the brain and the social milieu in which an individual develops. Theory of mind may be a cultural entity whose form gives it an efficient foothold in the neural processes of the human brain, to use an analogy to the "parasite" model proposed by Terrence Deacon (1997) with respect to language and the brain.

An adequate account of theory of mind's brain basis will have to explain varying degrees of impairment of theory of mind seen in clinical syndromes. Theory of mind has simple and subtle forms. The ability to understand that someone else has a false belief is in place by around age six in normal children; it is not until a number of years later that they acquire an understanding of faux pas, that is, that someone has said something they should not because they don't realize it produces a negative feeling state in someone else. This subtle form of theory of mind is impaired in people with the mild form of autism known as Asperger's syndrome; it is also impaired in people who have sustained injury to the medial prefrontal cortex (Stone et al. 1998).

Finally, at the far reaches of what we understand about the brain and social cognition is a recent discovery regarding the understanding of social exchange. Valerie Stone (Stone et al. 1997) tested a patient with a diffuse brain injury involving orbitofrontal cortex and the anterior temporal cortex, with consequent disconnection of the amygdala, all damage being bilateral. It had previously been determined that he had difficulty attributing mental states to others. In this study, Stone presented logical problems of two kinds. One involved using a precaution rule of the following type: "If you work with toxic chemicals, then you have to wear a safety mask." The other involved using a social contract rule: "If you borrow the car, then you have to fill up the tank with gas." Normal controls and subjects with brain damage involving different areas reason equally well with either type of rule. Stone's patient, however, performed significantly worse on social contract reasoning. Such a dissociation implies a separate brain mechanism for the two kinds of reasoning. However, as with theory of mind, there is not enough information to be able to specify the neural substrate of social-contract reasoning with any precision.

## CONCLUSION

Social stimuli have physical effects on neurons. Brains and the bodily processes they set into motion register such social signals as gaze direction, tone of voice, expressive movements of bodies, and movements of mouths and faces. Neural activity and bodily responses are the building blocks of our participation in the network of social acts. Since it is by virtue of social participation that the practices constituting mind emerge, realism about the mind belongs in the web of social signalling, not in socially constructed entities such as emotion, memory, rationality, and so forth.

# 10

## Look Again

### REVIEWING THE OPTIONS

We've seen that the neuroist approach to the mind-brain problem tries to fuse two separate discourses—one pertaining to the mind, the other to the material world—in a single picture. The results are awkward, as both neuroist documents and a persisting body of philosophical conundrums attest. We also saw that abandoning the neuroist strategy doesn't leave us with two separate worlds, one material and one mental. Because the brain-and-body is socially responsive in its makeup, human actions—always intrinsically embedded in systems of reasons and beliefs—do not float free from physical mechanisms.

One of the reasons philosophers have been stuck for so long in the mind-brain problem is that they tend to see the mental as a property of the individual—the mind is inside the person, so to speak. Some have tried to find better answers to the mind-brain dilemma through an externalist view (Heil 1992). Externalism holds that the mind has to be conceived more broadly: its functions are always in some context. As an analogy, we could not understand the ways that a quarter is used if we looked only at its molecular composition. To understand why quarters move about as they do, we have to place them in the context of an economic system. Absent the larger context, we could only understand low-level properties of the coin, like weight and hardness.

Likewise, says the externalist, how the nervous system translates chemical impulses into the contractile forces that move muscles and limbs is a restricted aspect of the body. The relation between brain activity and a raised arm is just a matter of physical events: like the relation between a brick and a brick wall, there is no discontinuity in getting from one level of description (neurons) to the other (limb motion). But getting from a physical sequence of causes, to my raising my arm to signal that I want to speak, to wave goodbye, or to command the attention of an orchestra—all these require a larger context *external* to my movement. An external account of mind means an act, or a word, takes its meanings from the uses to which it is put in social forms of life. Externalism goes in the right direction.

We'd have to use externalism to explain the relation between isolated bricks, and walls that function as property line markers, or as aesthetic decorations, or as blank slates for boys with spray paint cans. Just as in the case of human physical actions and mind, we'd be placing the bricks and the wall into a larger context, the context of human affairs. Similarly, the organized network of meanings, values, reasons and so forth that constitute the social world, is external to each of us. At the same time, it is composed of nothing more than a vast system of common gestures—from nods, glances, and grunts to inauguration speeches and contracts—which are, at root, material processes engaging bodies and brains. My bodily response to your raised eyebrows is material; my subsequent, elaborated act selects and creates a particular, intelligible sequence from the vast array of potential sequences that constitute our shared cultural resources.

Those who do not favor externalism—because they mistakenly see it as dualist—and are unhappy with the paradoxes resulting from internalism sometimes embrace a third alternative. They deny the mental, so that only the picture of the physical is left. Eliminativists— as those who make this choice are called—essentially assert that our ordinary grammar of person is mistaken. They believe it will one day be replaced by the language of neuroscience. (They assume that neuroscience will reverse its present subordination to psychology.)

Now imagine that the eliminativist vision has come to pass. The language of neuroscience has taken over, and neuroscientists have schooled the rest of us to speak about ourselves and our actions in new, more correct ways based on neural processes. The old concept of person that combined a body and a mental life has been laid to rest, and we no longer use it in our language or other interactions.

The minute we imagine such developments, however, we can see that eliminating talk of the mental simply pushes the paradoxes away from our theories and out into our behavior. To give just a single example, a sentence like "Behavior has only physical causes" would negate its own content, because uttering it signals the intention (a mental state) to communicate something (Malcolm 1968). This kind of paradox suggests that if we *could* banish the language of mind and the behavior that goes with it, the things we'd be able to say and do would be so different we wouldn't recognize ourselves as the same species. Changing our ordinary grammar of mind would be changing who we *are*. So to say our ordinary ways of talking are "mistaken" can't be right: unlike beliefs that we can change without doing violence to our own natures, our notions of person and mind penetrate our lives and actions through and through.

In fact, externalism also does some slight violence to our ordinary ways of speaking of persons, for we ordinarily speak and act as though the mind is inside the person, not a matter of social context. Eliminativism is much more extreme, for it aims to wipe out ordinary person language completely. Neither of them fully respects our ordinary ways of talking. But these ways must be respected, for they determine what we are.

This brings us to ordinary language philosophy. It is not much in fashion today as a route to understanding the mind. For one thing, ordinary language traditions have been derided because they are apparently cut off from neuroscience and the excitement it generates. Second, they are externalist because they situate meaning in social practices, and externalism slightly strains our ordinary way of talking about mind and person, whose flavor is internalist. Nevertheless, the ordinary language stance can take us in the direction we want. It has to do two things at once: it must respect the everyday, internalist ways we use the concept of person, and simultaneously recognize that it is those selfsame practices that *determine who we are*. It has to be externalist about internalism, in sum.

## THE SOCIAL DIMENSION

Ordinary language philosophy takes seriously a dimension of life we usually ignore. It is the dimension of social forms of life. We can think of it as the medium that produces human beings by virtue of their participation in it. They participate through adopting some subset of the myriad of specific social forms—modes of dress, forms of speech, objects arranged around them, activities carried out—available from the culture. This social dimension is the defining feature of the natural ecology of human beings, as water is the defining element of the natural ecology of fish.

The human individual doesn't really exist as a person until he or she takes up and participates in forms of social life. Perhaps this reflecting emptiness is what Shakespeare had in mind when he wrote that man is *"Most ignorant of what he's most assur'd, | His glassy essence"* (*Measure for Measure* 2.2.121–22). The phrase "glassy essence" was used extensively by the philosopher Charles Sanders Peirce, whose idea it was that "man is a symbol" (Singer 1980).

Our only reality is that we adopt social forms: we participate. Imagine some huge supergame, like Monopoly and other games combined, but infinitely more complex. To be human is to be one of the players, to follow the rules. Players can offer innovations in the rules; if enough people agree, the rules can be changed or new rules added, so the form of the game keeps evolving. But no one can be outside it while still experiencing himself or herself in the normal way, as a self (subject) among other selves.

The mistake of the internalists is to think that a behavior I may display, such as moving my piece counterclockwise, is an internal property of me. The externalist, by contrast, would see it as a property of my physical self and the context together, where the context is the series of possible locations of the piece, the history of its previous locations, the locations of other pieces, and so on. The clever ordinary language philosopher would see all the externalist sees, especially the arbitrary, social nature of the actual activity, and something more. That something is the fact that my own view of myself—and others' views of themselves—is that we *are* "counterclockwise piece movers," not players of a game with arbitrary rules. No one could tell me that I am *really just* a player of a game: *"just"* according to what vantage point? Our ways of talking about us *are* us. The game is all there is.

Being a person, with the dual aspects of body and mind, is following a set of rules we all learn (except under very unusual circumstances). As players of this game, we *really* have minds. From an outside point of view, though, our minds are *socially created* things we enact and talk about in rule-governed ways. The fact that the person concept is so widespread culturally, and persistent historically, should not tempt us to say it must be internal. A rule will be widespread and persistent to the extent that it stabilizes games—forms of life—and allows them to continue. Person rules apparently are like that.[1]

## HIDDEN PITFALLS

Psychology is a formal way of talking about, and therefore inventing, person rules. What makes psychology both true and real in a particularly interesting way is that we *are* the stories we tell about ourselves. Every psychological theory is true, if we take it up and enact it; and psychology is real to the extent that we make it part of our forms of life and act according to our understanding of it. Given the potency of psychology, we have to be careful:

> [P]sychologists come to argue, not just that we should think about ourselves and others in new ways, but that we should treat one another in new ways because they are ways with a scientifically established reality to them. . . . And this may happen as a result of work in different spheres of psychology irrespective of whether the results in those spheres are of scientifically satisfactory character or not. It can happen just as a result of suggesting that men should be thought of as being entities of one kind rather than another in order to investigate them in a scientific manner. "We are what we pretend to be," says Kurt Vonnegut, Jr., "So we must be careful about what we pretend to be." (Shotter 1975, 28)

We may have no option but to make up stories about ourselves and then live them, but we still have to take responsibility for them. Neuroism is a recent and popular kind of psychological story, but its implications have to be carefully examined. For example, shall we treat one another and ourselves as walking brains, attributing our behavior solely to good or bad neurotransmitters? Shall we acquiesce in society's attempts to change people's behavior by applying neuroactive chemicals and seeking the genes that produce faulty brains?

In general, there are two ways in which the brain metaphor can be harmful. One is that it might lead us to focus on neural tissue as the cause of our collective tensions and failings, instead of looking at our social organization and how it distributes resources of all kinds, including subsistence means, knowledge, and opportunities for participation. This narrowing and diverting of our attention may allow certain economic and social arrangements to be perpetuated even though they are unjust or harmful.

The second is that neuroism defines us in terms of something—the brain—that is only accessible through the mediation of an expert class. In effect, our own natures do not belong to us but to someone else. The experts are something like the officials of the church in the Middle Ages: in those times, one's soul was managed and maintained only through participation in the church, via contact with its clergy, and according to theories designed by its theologians. The Reformation, of course, sought to put the individual more directly in charge of his or her own soul. Neuroism is akin to a religion in which the essential control of the self—knowledge about it, means for changing it—belongs to an elite class of brain experts, rather than to everyday individuals.

Our neural machinery produces organized patterns of physical activity by which we participate in social forms of life. It doesn't produce mind; it enables participation. But as we've seen, one of the most important social forms in which individuals can participate is "person"—which is where mind is to be found. Therefore, it is incorrect to look for any particular part of our vocabulary of mind—attention, memory, emotion, or even some modified forms of them—in brain processes. Instead, we must study the brain to understand further how it supports our participation in elaborate social forms, forms that include paying attention, remembering, and being emotional.

In conclusion, nothing was hidden behind or within the everyday picture of person. What was hidden was the social dimension that produced it. When we don't see that dimension, there is a pull toward neuroism, the result of taking our everyday internalist talk too seriously. As a science of mind, neuroism won't do. We have to look again.

# Notes

## 1. THE MIND-BRAIN PROBLEM

1. This capsule description of the problem and possible solutions is extremely condensed. The reader may consult standard reference works in philosophy or cognitive science for more detail.
2. A minority view is that advances in neuroscience aren't bringing us anywhere close to a solution. See Horgan (1999).

## 2. PICTURES

1. This is clearly explained in McGinn (1997, 147).
2. McGinn (1997) explains: "What Wittgenstein wants us to see is that this move from 'the body' to 'the subject who feels the pain' (to 'I') is not a movement between entities, but a *grammatical movement*, a movement between language games" (156).
3. This is the version of science adhered to by most scientists and laypeople, and therefore the version relevant to the dual picture we are about to examine. It is itself a set of social practices, as sociologists of science have demonstrated. Phrases concerning the language of science in this chapter refer to this set of practices, which has also been called "naïve realism." See Hesse (1980, vii).
4. For detailed demonstrations, see Gergen (1987, 115–29).

## 3. PROBLEM? WHAT PROBLEM?

1. The citation is irrelevant in any case, for the original quote refers to the dorsolateral, not the orbitofrontal, cortex.
2. True, neuroscientists have proposed that the prefrontal cortex allocates attention, acting as a sort of "executive" that chooses among competing tasks, an activity which could be expressed somewhat loosely as "weighing actions." However, recent evidence (Adcock et al., 2000) has thrown this interpretation

of prefrontal function into doubt, illustrating once again how tentative our understanding of the brain actually is.

## 4. Bringing out the hammers and saws

1. Neuroists are also haunted by Freud, as we shall see, because of the way that Freud stands, in the popular mind, for truth and authority in narratives of the mental.
2. I have argued elsewhere (Brothers 1997) that this is exactly where emotion belongs—in our everyday narratives and performances, not in our brains.
3. The nature of Solms's internal sensory organ is vague; it seems to point, like a telescope or camera: "The supposedly mysterious distinction between mind and brain thus dissolves into a simple distinction between different perceptual modalities, facing in different directions" (Solms 1997, 700). What does "facing in different directions" mean? Unless the internal sensor is a neural entity, it is a homunculus seated in a Cartesian theater. The key to the confusion may reside in the language of introspection: Wittgenstein pointed out that the grammar of reporting an experience makes it seem as though we "look inside."
4. It is left to the reader of the sweepstakes mailings to figure out, from still finer print, that the chances are overwhelming that he or she will never be a winner. The present work highlights neuroism's fine print message.
5. It is difficult to determine whether speculation is even viewed as problematic by neuroists. In some cases, it is portrayed as valuable. For example, Turner (2000) writes, "In chapter 3, I admitted to engaging in wide-eyed speculation, a danger that has only been compounded, of course, in this chapter. There are no data to bring to bear on the speculations here, but I believe that they are worth making, if only to stimulate thought and research on the topic" (145–46).

   Speculation is easy and the number of possible speculative narratives is almost infinite; these considerations should temper the urge to speculate neuroistically. However, they do not, which suggests that any embarrassment is overridden by other factors. In a later chapter, we will suggest that these factors are social motivations and interests.

## 6. Getting it wrong

1. Refreshingly, this issue of *Zygon* included some critical analyses of the d'Aquili and Newburg effort. These not only pointed out the weak scientific bases for d'Aquili and Newburg's claims, but also commented on the general implications of neuroscience's incompleteness. See Holmes (1993).

## 7. Cults, mysteries, and money

1. This has been asserted by E. O. Wilson (1998).
2. To argue against eliminative reduction, say some proponents, is to argue based on what we do not know, since eliminative reduction *could* provide the

answer (we just don't know what it is yet). This is a rather weak logical argument, equivalent to saying, "We have faith in the future of brain research." Faith is a moral, not a logical position. What is being held at bay is the devil of dualism.

3. Groups and institutions that should be representing the public's interest in the scientific objectivity of the information these individuals disseminate fail to do so. Among those who should require disclosure of the income certain psychiatrists receive from pharmaceutical companies, but do not, are editorial boards of the journals in which their psychopharmacology research studies are published; the accrediting agencies for graduate (residency) education in psychiatry, as regards the faculty of the programs they accredit; and academic departments, which bestow titles such as "professor" that imply academic disinterest. Furthermore, states require that physicians attend a certain number of courses throughout their careers in order to stay current in their field. These requirements are called "continuing medical education" or CME. Commercial providers of CME in psychiatry are often thinly veiled fronts for pharmaceutical companies which offer their courses free of charge in luxurious settings. The American Council of Continuing Medical Education, which accredits the providers of such courses, has guidelines regarding the information disseminated and the commercial ties of the presenters that practically speaking do little to prevent penetration by commercial interests.

4. The question "Are my symptoms due to a chemical imbalance?" is different from the question, "Are my symptoms likely to improve if I take a psychoactive medication?" Clinicians know that drugs such as antipsychotics and serotonin reuptake inhibitors can improve certain symptoms, but no one knows how or why they work. (To know this, one would have to understand how the brain works.) The chemical imbalance concept is a post hoc explanation, not a demonstrated entity.

5. Bloom's position ignores the fact that psychiatric diagnoses are not objective entities at all; in fact, the "rules" for making a diagnosis are codified through the negotiations of small committees of academic psychiatrists. Neuroscientists are not practicing psychiatric clinicians, and it is unsurprising that they are less critical of official psychiatric diagnoses than those who wrestle with the complexities of real human beings as part of their daily practice, who have seen fashions of psychiatric diagnosis come and go, and who are aware of the political nature of the process whereby the standard diagnostic handbooks are drawn up. In point of fact, psychiatric diagnoses are socially constructed entities molded by psychiatric politics and covert pharmaceutical interests. They therefore cannot provide an independent framework for neuroscience research.

6. Because of the secrecy surrounding the industry's payments to individual psychiatrists, and the absence of public records which would reveal a researcher's sources of income, the dollar amounts of industry support for Gorman and others like him are not known. However, as of May 1999, Gorman was receiving grants and research support from Eli Lilly and Company and SmithKline Beecham Pharmaceuticals. He had received honoraria from Eli Lilly and Company, Forest Laboratories, Janssen Pharmaceutica, Inc.,

Lundbeck, Merck & Co., Inc., Organon Inc., Parke-Davis, Pfizer, Inc., SmithKline Beecham Pharmaceuticals, Wyeth-Ayerst Laboratories, and Zeneca Pharmaceuticals (information from required disclosure to CME, Inc.).

## 8. AN UNEQUAL MATCH

1. Pribram (1980) explicitly advocates the use of technological analogies, such as holograms and others mentioned here, for understanding the brain. See Stoljar and Gold (1998) for additional discussion.

## 9. SOCIAL NEUROSCIENCE

1. This was discovered by Johansson (1973).
2. Recently, the use of the term "amygdala" has been questioned due to the heterogeneity of the structures to which it refers. See Swanson and Petrovich (1999).

## 10. LOOK AGAIN

1. I've suggested elsewhere (Brothers 1997) that these rules, and the human brain, may have evolved cooperatively. This would give the rules considerable cultural stability—just like language. If the idea is correct, it would explain why some kinds of brain disorder cause breakdown in an individual's ability to follow person rules.

# References

Adcock, R., R. Constable, J. Gore, and P. Goldman-Rakic. 2000. Functional neuro-anatomy of executive processes involved in dual-task performance. *Proc. Nat. Acad. Science* 97: 3567–72.

Adolphs, R., D. Tranel, H. Damasio and A. Damasio. 1994. Impaired recognition of emotion in facial expressions following bilateral damage to the human amygdala. *Nature* 372: 669–72.

Adolphs, R. 1999. Social cognition and the human brain. *Trends in the Cognitive Sciences* 3: 469–79.

Austin, J. 1998. *Zen and the brain: Toward an understanding of meditation and consciousness.* Cambridge, Mass.: MIT Press.

Belin, P., R. Zatorre, P. Lafaille, P. Ahad, and B. Pike. 2000. Voice-selective areas in human superior temporal sulcus. *Cognitive Neuroscience Society Abstracts* 63A, 31.

Bentin S., T. Allison, A. Puce, E. Perez, and G. McCarthy. 1996. Electrophysiological studies of face perception in humans. *J. Cognitive Neuroscience,* 8: 551–65.

Bernstein, L., E. Auer, J. Moore, C. Ponton, M. Don, and M. Singh. 2000. Does auditory cortex listen to visible speech? *Cognitive Neuroscience Society Abstracts* 99A, 37.

Bloom, F. 1995. Cellular mechanisms active in emotion. In *The Cognitive Neurosciences,* ed. M. Gazzaniga. Cambridge, Mass.: MIT Press, 1063–70.

Bonda, E., M. Petrides, D. Ostry, and A. Evans. 1996. *J. Neuroscience* 6: 3737–44.

Broadbent, D. E. 1958. *Perception and communication.* Exeter, U.K.: A. Wheaton and Co.

Brothers, L. 1990. The social brain: A project for integrating primate behavior and neurophysiology in a new domain. *Concepts in Neuroscience* 1: 27–51.

———. 1997. *Friday's footprint: How society shapes the human mind.* New York: Oxford University Press.

Brothers, L., and B. Ring. 1992. A neuroethological framework for the representation of minds. *J. Cognitive Neuroscience* 4: 107–18.

Cabeza, R., and L. Nyberg. 2000. Imaging cognition II: An empirical review of 275 PET and fMRI studies. *J. Cognitive Neuroscience* 12: 1–47.

Collins, R. 1989. Toward a neo-Meadian sociology of mind. *Symbolic Interaction* 12: 1–32.

Damasio, A. 1994. *Descartes' error: Emotion, reason, and the human brain.* New York: G. P. Putnam's Sons.

d'Aquili, E., and A. Newberg. 1993. Religious and mystical states: A neuropsychological model. *Zygon* 28: 177–200.

Deacon, T. 1997. *The symbolic species: The co-evolution of language and the brain.* New York: W. W. Norton.

Dennett, D. 1991. *Consciousness explained.* Boston: Little, Brown and Co.

Durkheim, E. [1912] 1995. The elementary forms of religious life. Trans. Karen E. Fields. New York: Free Press.

Engel, A., P. König, A. Kreiter, T. Schillen, and W. Singer. 1992. Temporal coding in the visual cortex: New vistas on integration in the nervous system. *Trends in Neuroscience* 15: 218–26.

Fuster, J. 1991. The prefrontal cortex and its relation to behavior. *Progress in Brain Research* 87: 201–11.

Gazzaniga, M. 1995. Preface in *The cognitive neurosciences.* Cambridge, Mass.: MIT Press, xiii-xiv.

Gergen, Kenneth. 1987. The language of psychological understanding. In *The analysis of psychological theory: Metapsychological perspectives,* ed. H. Stam, T. Rogers, and K. Gergen. New York: Hemisphere Publishing.

Gloor, P. 1986. The role of the human limbic system in perception, memory, and affect: Lessons from temporal lobe epilepsy. In *The limbic system: Functional organization and clinical disorders,* ed. B. K. Doane and K. E. Livingston. New York: Raven Press, 159–69.

Gold, I., and D. Stoljar. 1999. A neuron doctrine in the philosophy of neuroscience. *Behavioral and Brain Sciences* 22: 809–30.

Goleman, D. 1994. *Emotional intelligence: Why it can matter more than IQ.* New York: Bantam Books.

Griffiths, P. 1997. *What emotions really are: The problem of psychological categories.* Chicago: University of Chicago Press.

Grotstein, J. 1994. Foreword to *Affect regulation and the origin of the self: The neurobiology of emotional development,* by A. Schore. Hillsdale, N.J.: Lawrence Erlbaum Associates, xxi–xxvii.

Healy, D. 1997. *The antidepressant era.* Cambridge, Mass.: Harvard University Press.

Hebb, D. O. 1949. *The organization of behavior: A neuropsychological theory.* New York: John Wiley and Sons.

Heil, J. 1992. *The nature of true minds.* Cambridge: Cambridge University Press.

Hesse, M. 1980. *Revolutions and reconstructions in the philosophy of science.* Bloomington: Indiana University Press.

Hoffman, E., and J. Haxby. 2000. Distinct representations of eye gaze and identity in the distributed human neural system for face perception. *Nature Neuroscience* 3: 80–84.

Holmes, H. R. 1993. Thinking about religion and experiencing the brain: Eugene d'Aquili's biogenetic structural theory of absolute unitary being. *Zygon* 28: 201–15.

Horgan, J. 1999. *The undiscovered mind: How the human brain defies replication, medication, and explanation.* New York: Free Press.

Humphrey, N. 1983. *Consciousness regained: Chapters in the development of mind.* New York: Oxford University Press.

Johansson, G. 1973. Visual perception of biological motion and a model of its analysis. *Percept. Psychophys.* 14: 202–11.

Joseph, R. 1990. *Neuropsychology, Neuropsychiatry, and Behavioral Neurology.* New York: Plenum Press.

Kanwisher, N., J. McDermott, and M. Chun. 1997. The fusiform face area: A module in human extrastriate cortex specialized for face perception. *J. Neuroscience* 17: 4302–11.

Karmiloff-Smith, A., E. Klima, U. Bellugi, J. Grant, and S. Baron-Cohen. 1995. Is there a social module? Language, face processing, and theory of mind in individuals with Williams syndrome. *J. Cognitive Neuroscience* 7: 196–208.

Kling, A., and H. Steklis. 1976. A neural substrate for affiliative behavior in nonhuman primates. *Brain, Behavior and Evolution* 13: 216–38.

Klüver, H., and P. Bucy. 1937. "Psychic blindness" and other symptoms following bilateral temporal lobectomy in rhesus monkeys. *Amer. J. Physiology,* 119: 352–53.

LaNoue, M., C. Edgar, and M. Weisend. 2000. Magnetoencephalographic correlates of the processing of emotional facial expressions. *Cognitive Neuroscience Society Abstracts* 27C, 70.

LeDoux, J. 1991. Emotion and the limbic system concept. *Concepts Neurosci. (World Scientific)* 2(2): 169–99.

———. 1995. Introduction to emotion. In *The cognitive neurosciences,* ed. M Gazzaniga. Cambridge, Mass.: MIT Press, 1047–48.

———. 1996. *The emotional brain: The mysterious underpinnings of emotional life.* New York: Simon and Schuster.

Maddox, J. 1998. *What remains to be discovered: Mapping the secrets of the universe, the origins of life, and the future of the human race.* New York: Free Press.

Magoun, H. 1964. *The waking brain.* Springfield, Ill.: Charles C. Thomas.

Malcolm, N. 1968. The conceivability of mechanism. *Philosophical Rev.* 77: 45–72.

McGinn, M. 1997. *Routledge philosophy guidebook to Wittgenstein and the philosophical investigations.* London: Routledge.

Merin, N. 2000. Gaze processing deficits in Fragile X syndrome investigated using fMRI. *Cognitive Neuroscience Society Abstracts,* 78B, 57.

O'Craven, K., and N. Kanwisher. 2000. Mental imagery of faces and places activates corresponding stimulus-specific brain regions. *J. Cognitive Neuroscience* 12: 1013–23.

Pally, R. 1998a. Emotional processing: The mind-body connection. *Int. J. Psychoanalysis* 79: 349–62.

———. 1998b. Bilaterality: Hemispheric specialisation and integration. *Int. J. Psycho-analysis* 79: 565–78.

Pally, R., and D. Olds. 1998. Consciousness: A neuroscience perspective. *Int. J. Psycho-analysis* 79: 971–89.

Perkel, D. 1987. Chaos in brains: fad or insight? *Behavioral and Brain Sciences.* 10: 180–81.

Persinger, M. 1996. Feelings of past lives as expected perturbations within the neurocognitive processes that generate the sense of self: Contributions from limbic lability and vectorial hemisphericity. *Perceptual and Motor Skills* 83: 1107–21.

Pert, C. 1997. *Molecules of emotion:Why you feel the way you feel*. New York: Scribner.

Pribram, K. 1980. The role of analogy in transcending limits in the brain sciences. *Daedelus* 109(2): 19–38.

Price, J. L., S. T. Carmichael, and W. C. Drevets. 1996. Networks related to the orbital and medial prefrontal cortex: A substrate for emotional behavior? In *Progress in Brain Research*. Vol. 107, ed. G. Holstege, R. Bundler, and D. B. Saper. New York: Elsevier.

Puce, A., T. Allison, J. Gore, and G. McCarthy. 1995. Face-sensitive regions in human extrastriate cortex studied by functional MRI. *J. Neurophysiology* 74: 1192–99.

Puce, A, T. Allison, S. Spencer, D. Spencer, and G. McCarthy. 1997. Comparison of cortical activation evoked by faces measured by intracranial field potentials and functional MRI: Two case studies. *Human Brain Mapping* 5: 298–305.

Ramachandran, V. S., and S. Blakeslee. 1998. *Phantoms in the brain: Probing the mysteries of the human mind*. New York: William Morrow and Co.

Rapoport, J., and A. Fiske. 1998. The new biology of obsessive-compulsive disorder: Implications for evolutionary psychology. *Perspectives in Biology and Medicine* 41: 159–75.

Restivo, S. 1985. *The social relations of physics, mysticism, and mathematics*. Dordrecht: D. Reidel.

Rouse, J. 1996. *Engaging science: How to understand its practices philosophically*. Ithaca: Cornell University Press.

Sacks, O. 1985. *The man who mistook his wife for a hat and other clinical tales*. New York: Summit Books.

Schore, A. 1994. *Affect regulation and the origin of the self: The neurobiology of emotional development*. Hillsdale, N.J.: Lawrence Erlbaum Associates.

Schwartz, J. 1999. First steps toward a theory of mental force: PET imaging of systematic cerebral changes after psychological treatment of obsessive-compulsive disorder. In *Toward a science of consciousness 3: The third Tucson discussions and debates,* ed. S. Hameroff, A. Kaszniak, and D. Chalmers. Cambridge, Mass.: MIT Press, 111–22.

Scott, S., A. Young, A. Calder, D. Hellawell, J. Aggleton, and M. Johnson. 1997. Impaired auditory recognition of fear and anger following bilateral amygdala lesions. *Nature* 385: 254–57.

Searle, J. 1997. *The mystery of consciousness*. New York: New York Review.

Shotter, J. 1975. *Images of man in psychological research*. London: Methuen.

Singer, M. 1980. Signs of the self: An exploration in semiotic anthropology. *Amer. Anthropol.* 82: 485–507.

Skarda, C., and W. Freeman. 1987. How brains make chaos in order to make sense of the world. *Behavioral and Brain Sciences* 10: 161–95.

Solms, M. 1997. What is consciousness? *J. American Psychoanalytic Assoc.* 45: 681–703.

Stephan, H., H. Frahm, and G. Baron. 1987. Comparison of brain structure volumes in insectivora and primates 7. Amygdaloid components. *Journal für Hirnforschung* 28: 571–84.

Stoljar, D., and I. Gold. 1998. On biological and cognitive neuroscience. *Mind and Language* 13: 110–31.

Stone, V. E., S. Baron-Cohen, L. Cosmides, J. Tooby, and R. T. Knight. 1997. Selective impairment of social inferences following orbitofrontal cortex damage. *Proceedings of the Nineteenth Annual Meeting of the Cognitive Science Society*, ed. Michael Shaft and Pat Langley. Mahwah, N.J.: Lawrence Erlbaum Associates.

Stone, V., S. Baron-Cohen, R. Knight. 1998. Frontal lobe contributions to theory of mind. *J. Cog. Neuroscience* 10: 640–56.

Strawson, P. 1959. Persons. In *Individuals: An essay in descriptive metaphysics*, 87–116. London: Methuen and Co.

Swanson, L. and G. Petrovich. 1999. What is the amygdala? *Trends in Neuroscience* 21: 323–31.

Thom, R. 1987. Chaos can be overplayed. *Behavioral and Brain Sciences* 10: 182–83.

Thomson, R. G. 1996. From wonder to error—a geneology of freak discourse in modernity. Introduction to *Freakery: Cultural spectacles of the extraordinary body*. New York: New York University Press, 1–19.

Tulving, E. 1995. Introduction [Memory Section] to *The cognitive neurosciences*, ed. M. Gazzaniga. Cambridge, Mass: MIT Press, 751–53.

Turner, J. 2000. *On the origins of human emotions: A sociological inquiry into the evolution of human affect*. Stanford: Stanford University Press.

*UCLA Medicine* 19, no. 2 (Fall 1998).

Velasco, M., and D. Lindsley. 1965. Role of orbital cortex in regulation of thalamocortical electrical activity. *Science* 149: 1375–77.

Wilson, E. O. 1998. *Consilience: The unity of knowledge*. New York: Alfred A. Knopf.

Wittgenstein, L. 1958. *Philosophical investigations*. 3d ed. Trans. G. E. M. Anscombe. Englewood Cliffs, N.J.: Prentice Hall.

Young, A., J. Aggleton, D. Hellawell, M. Johnson, P. Broks, and J. Hanley. 1995. Face processing impairments after amygdalotomy. *Brain* 118: 15–24.

# Index

Adolphs, R., 81, 84
American Psychiatric Association, 63
amygdala, 65, 96 n.2; ascription of
mental states and, 82–84; electrical
stimulation of, 82–83; emotion and,
19, 31; evolution of, 79, 81; fear
conditioning and, 31; lesions of, and
behavior, 53, 81–84; social signals
and, 78, 81
analysis, critical: discouragement of,
40; immunizing narratives against,
13, 41
antidepressants, 63, 65. *See also* drugs,
marketing of
anxiety. *See* disease, psychiatric
Asperger's syndrome, 85
astronomy, 69
athletic performance, and the brain, 5–
6
attractors, 70
audiences, 40, 45, 62
Austin, J., 40–41
autism, 78, 84, 85

behavior: and ordinary language philosophy, 2. *See also* Damasio; emotion, and behavior; qualia
Bentin, S., 79, 80
Bloom, F., 64
body: brain response to sight of, 81;
consciousness and, 15; in mind-
brain problem, 1; and person con-

cept, 9; role in feeling, 17–19; social
context and, 75, 83, 86, 88
Bonda, E., 80–81
brain: advertising images of, 5, 62;
decade of the, 6; evolution of, 78–
79; internalism and, 2; naturalization
and, 3; PET and MRI scans of, 16,
24, 79–82; "social," 78, 79; temporal
lobes of, 57, 77–80. *See also* cult, of
brain; disease, neurological; mind-
brain problem; mystical states; neuroscience, absence of central theory
in; sacred object, brain as
Broadbent, D. E., 70
Brothers-Ring hypothesis, 84–85

Cabeza, R., 68
chaos theory, 70–71
"chemical imbalance," 63, 95 n.4
children: self-calming in, 14 self-
control in, 15, 47
Christianity, 59, 60
citations, of scientific articles, 14, 15,
23, 47, 50–51
cognition, social, 76–86
Collins, R., 60
complexity. *See* neuroism, and
complexity
concepts. *See* part-concept
consciousness. *See* unconscious
contexts: psychological versus neuroscientific, 13, 14–16; and mind, 87–88